RUSSIAN

SPACECRAFT

by Robert Godwin

Acknowledgements

This book would not have been possible without the sterling work of Asif
Siddiqi, Mark Wade, Frank Winter and Anatoly Zak whose books and web-
sites provided the vast majority of the insights contained herein. However,
any errors that may have crept into this text are undoubtedly my own and
are in no way attributable to this group of outstanding experts.

We acknowledge the financial support of the Government of Canada through the Book Publishing Industry Development Program for our publishing activities.

Published by Apogee Books, Box 62034, Burlington,
Ontario, Canada, L7R 4K2, http://www.apogeebooks.com
Tel: 905 637 5737
Printed and bound in Canada
Russian Spacecraft by Robert Godwin
ISBN.   1-894959-39-6
ISBN 13.   9781-894959-39-1
©2006 Robert Godwin
Images are courtesy NASA and Energia except where noted.

Introduction

During the last five decades the great majority of Russian spacecraft have been pushed into space by variations of two launch vehicles. For the purposes of this book they are referred to as the *R-7* and the *UR500/Proton*. Although there have been others, these two designs have proven to be so tough and versatile that they have carried the lion's share of the payloads.

The R-7 has been the manned launch vehicle since 1961 and has been lofting unmanned payloads since 1957, while the UR500/Proton has been the principal unmanned heavy launcher since 1965. The R-7 has undergone at least four major upgrades and countless minor revisions. The first upgrade added a third stage which sent *Luna 1* to the moon and lofted Yuri Gagarin's *Vostok* into the history books. The next upgrade turned it into the *Voskhod* launcher that carried the first three-man crew. Mars and Venus probes required the addition of a fourth stage and so the R-7 became known as the *Molniya*. Finally, a fourth major overhaul turned it into the *Soyuz* and *Progress* launcher used today. The R-7 was often renamed after each upgrade while, despite the addition of a variety of upper stages and payloads, the UR500 has remained the *Proton* since its first space launch.

The reader may also be confused by the tendency to rename spacecraft. The generic term *Cosmos* was used for hundreds of satellites and probes but if the spacecraft performed some important historic function it would often be renumbered and renamed.

Navigating the labyrinth of people, places and organizations that comprised the world's first space program is a task that would easily have confounded Theseus. Compounding the problem is the Russian propensity for renaming things and shifting responsibilities from one factory to another. The names of the factories, spacecraft, rockets, launch facilities and even towns have changed, and all of this is further complicated by most of the story taking place behind a wall of secrecy that lasted for nearly forty years. Rather than confuse the reader with the constantly shifting dominions of the various design bureaus, some names have been simplified or deliberately omitted. The next four pages provide an overview.

Principal players:

Vladimir Chelomei (1914 – 1984)
  Architect of Proton rocket, space stations, winged spacecraft
Valentin Glushko (1908 – 1989)
  Principal Russian rocket engine designer
Helmut Grottrup (1916-1981)
  Captured German V2 rocket scientist
Alexei Isayev (1908 – 1971)
  Rocket engine designer
Sergei Korolev (1906 – 1966)
  Architect of R-7 rocket, Vostok, Sputnik, Soyuz etc.
Nikolai Kuznetsov (1911 – 1995)
  Rocket engine designer
Semyon Kosberg (1903 – 1965)
  Rocket engine designer
Vasiliy Mishin (1917 – 2001)
  Successor and assistant to Korolev at OKB-1
Mikhail Tikhonravov (1901 – 1974)
  Theoretician and designer for early Soviet spacecraft
Friedrich Tsander (1887 – 1933)
  Builder of first Russian non-solid rocket
Dimitri Ustinov (1908 – 1984)
  Minister of Armaments and political overseer to Soviet missile program
Mikhail Yangel (1911 – 1971)
  Architect of Soviet storable propellant missiles and space launchers

Important places:

Kapustin Yar
  Rocket launch facility between Volgograd and the Kazakhstan border
Baikonur
  Principal Soviet space launch facility located near to Tyuratam in Kazakhstan
OKB-1
  Korolev's design bureau, later renamed TsKBEM, and Energia
OKB-456
  Glushko's design bureau, later part of Energia, then renamed Energomash
OKB-586
  Yangel's design bureau, later renamed Yuzhnoye
OKB-52
  Chelomei's design bureau, today known as NPO Mashinostroyenia
OKB-301
  Lavotchkin design bureau
OKB-276
  Kuznetsov's design bureau
OKB-154
  Kosberg's design bureau
Molniya
  A division of the MiG bureau and builder of the Buran shuttle (also see†)
Khrunichev Machine Building Plant
  Was part of OKB-52, then combined with Salyut bureau, now makes space station modules and Proton launch vehicles.

Important spacecraft:

Almaz – Chelomei's military space station and observation platform
BOR-4 – unmanned winged re-entry test vehicle (half scale Spiral)
BOR-5 – unmanned winged re-entry test vehicle (one eighth scale Buran)
Buran – Soviet space shuttle
FGB – Cargo unit for Almaz, used as basis for small Mir and ISS modules
L1 – Korolev's proposed manned lunar orbiting aggregate
L3 – Korolev's proposed manned lunar landing aggregate
LK – Korolev's proposed manned lunar lander, part of L3
LK-1 – Chelomei's proposed manned lunar orbiter
LOK – Korolev's proposed manned lunar mothership, part of L3
Luna – unmanned robotic lunar orbiters and landers (also known as Lunik)
Lunokhod – unmanned lunar rover, also known as E-8
Mars – unmanned Mars probe
Mir – large Soviet/Russian space station, main *Mir* module based on Almaz
Progress – modified Soyuz used for cargo transport
Salyut – space station based on Almaz, seven different Salyut flew
Soyuz – two or three man spacecraft also known as 7K
Spiral – Proposed winged shuttle
Sputnik – The first earth satellite, also known as PS-1
Sputnik 2 – The second earth satellite, carried dog Laika
Sputnik 3 – Third earth satellite, also known as Object-D
T2K – unmanned, legless version of LK lunar lander
TKS – Chelomei's proposed supply vessel for Almaz
VA (Merkur) – Chelomei's proposed re-entry vehicle for Almaz
Venera – unmanned Venus probe
Voskhod – converted Vostok to carry extra passengers
Vostok – First manned spacecraft, also known as Object-K
Zenit – unmanned Vostok, converted to become a spy satellite (also see*)
Zond – modified unmanned LOK (also name for unrelated planetary probes)

Important launch vehicles:

Energia – Glushko's super heavy-lift cryogenic two stage booster for Buran
Kosmos – Yangel's small satellite launcher modified from the R-14 IRBM
Molniya – R-7 with I and L third and fourth stage upgrades (also see†)
N-1 – Korolev's giant booster for manned lunar landings
Proton – Chelomei's heavy lift booster based on UR-500 missile
R-1 – first Soviet version of the V2
R-14 – Yangel's storable propellant ICBM, became Kosmos space launcher
R-16 – Yangel's storable propellant ICBM
R-2 – upgraded Soviet version of the R-1
R-5 – upgraded Soviet missile, flew scientific payloads from 1953-64
R-7 – first space launcher, known as SS-6, Semyorka, CH-10 and Sapwood
R-9 – Korolev's smaller cryogenic ICBM
Soyuz – R-7 with I third stage upgrade and launch escape tower
Tsiklon – (Cyclone) Yangel's medium lift launcher based on R-36 ICBM
V2 – German ballistic missile also known as the A-4
Vostok – R-7 with E-1 third stage upgrade
Zenit – Large strap-on booster for Energia modified for satellite launches.*

Russian Spacecraft

It was Friedrich Tsander, in late 1931, who engineered the earliest success for the Moscow branch of the *Group to Study Reaction Propulsion* (GIRD), one of the first Russian "rocket clubs". Tsander was the originator of the OR-1, a primitive rocket engine that he had built in the late 1920's from a welding torch and a sparkplug. Much like the VfR in Germany, GIRD was a loosely based group of individuals spread across the Soviet Union. Its mandate, such as it was, involved aviation, gliding, chemical research and a host of other interests and disciplines. It formed when Tsander met Sergei Pavlovich Korolev in the summer of 1931. Since Tsander was not interested in being a manager of the organization, he left that task to his bright, young protegé. Korolev and Tsander had begun their early work in a wine-cellar, where they hosted the Moscow branch of GIRD. By May of 1932 Korolev replaced Tsander as the group's administrative leader.

Sergei Korolev

Running almost concurrently with GIRD was another group that was known as the GDL. The so-called *Gas Dynamics Laboratory* had evolved from a decades-old military facility whose principal work had been with solid fuel rockets, but the GDL also did research with the new liquid propellants. If GDL had any equivalent in the west it might be the US Army's *Jet Propulsion Laboratory* in Pasadena. JPL's earliest founders would later leave to found *Aerojet*, a formidable contractor and builder of rocket engines, while the GDL would do likewise when its most famous son, Valentin Glushko, would create Special Design Bureau (OKB) 456. It is difficult to find an obvious counterpart in the west for Glushko. His direct influence on engine designs would extend through most of the 20th century, beginning in the 1920's and culminating in the 1980's at which point he had merged his own engine design bureau (OKB-456) with the more illustrious OKB-1 (Korolev's bureau) to create the almost monolithic *Rocket & Space Corporation Energia*. After his death, the Glushko engine division would once more leave the Energia umbrella and revert to a stand-alone organization called NPO Energomash.

Valentin Glushko

GIRD's first liquid fuel rocket, incidentally *not* designed by Korolev, was launched on August 17th 1933. Many people in the West think that Korolev was an engineer and a scientist and that he had a hand in the machinations of designing most of the hardware. This probably stems from his title, which came later in life, of *Chief Designer*. Korolev was certainly a capable engineer, he had built and flown his own aircraft in his youth, but his greatest strength lay in his management skills. The comparisons to Wernher von Braun are well founded in that respect. Both men were brilliant managers, both were excellent strategists (an important asset when dealing with politicians) and both had a clear vision of what they wanted. However, whereas von Braun was, first and foremost, the father of the Redstone and Saturn launch vehicles, it would be doing a disservice to Korolev to suggest that he only produced a handful of boosters. He was the spiritual father of the R-7 and the giant N-1 lunar boosters *and* the spacecraft that flew on them. He was also the guiding hand behind many of the Soviets' robotic space vehicles. Another significant difference seems to be that von Braun had years of experience with engines, whereas Korolev was always at a disadvantage in that area, frequently requiring the assistance of Glushko, or of other less well-known engine designers, such as Alexei Isayev and Nikolai Kuznetsov.

Just one month after GIRD's first successful rocket flight the group consolidated operations with its erstwhile competitor the GDL, and the new hybrid became known as RNII. Unfortunately the GDL engineers had been working with solid propellants since the early 1920's and at first they didn't think much of the GIRD team, or their liquid fueled rockets. However, they gradually put their differences behind them and work continued for the next six years until the sadistic and paranoid psychotics under Stalin's command turned a baleful eye on the group. The Army officer who had helped bring GDL and GIRD together was one of the first casualties, he was tried and summarily executed on virtually no evidence of any crime. Inevitably, anyone who had spoken to him came under suspicion, including the young engineers at RNII. Two of the principal members of the group were arrested and subsequently tortured until, under great pressure, they falsely accused and denounced both Glushko and Korolev for anti-Soviet activities. Glushko was arrested in March 1938 and Korolev, hospitalized at the time, was detained three months later. Korolev was sentenced to ten years hard labor in one of the harshest prison camps in the Gulag, while Glushko was sent to serve eight years in a slightly less severe prison near Moscow.

Joseph Stalin's regime was one of the most callous, malicious and ultimately self-destructive governments in the history of the world. Millions of mostly innocent people were imprisoned or executed during his tenure. This paranoid megalomania seems to have come directly from Stalin himself and his instructions were viciously implemented by the security forces under his command. It was during Stalin's tenure that Soviet rocketry began to flourish, and so a cloak of secrecy and obfuscation was drawn around everything involving missiles. This mantle remained impenetrable for almost fifty years and effectively concealed from the world the true scale and scope of the Soviet space program.

After the end of the Cold war, and thanks to the era of glasnost, the actual structure of the Soviet space bureaucracy was finally revealed to the west, and in no better place than in the superb book *Challenge to Apollo* by space scholar Asif Siddiqi. In this book the intricate details of Soviet space policy are meticulously outlined and the veil of half-truths and political expediencies are removed to reveal a system completely alien to anything conceived in the West. The system which created Sputnik, Vostok, Voskhod, Lunokhod, Venera, Soyuz, Salyut and Mir and a host of other extraordinary space achievements was, like its Western counterpart, spawned by enthusiastic rocketeers in the 1930's, nourished and enriched by Nazi German plunder and then appropriated by the military establishment. But if these things sound familiar and recognizable, think again – the parallels are superficial.

Both the Soviets and the Americans recognized the military value of the astounding breakthroughs discovered in the summer of 1945 at Peenemünde, but while the different branches of the American military would briefly squabble over the use of these assets, the feuding came to an abrupt end in 1958 when President Eisenhower decreed the establishment of a civilian agency to oversee the course of America's space program. In Stalin's Russia no such decree was ever issued and the fledgling Soviet space program was doomed to remain under the jurisdiction of powerful military agencies. There would be no Russian equivalent to NASA (at least not until 1992 when Boris Yeltsin created the Russian Space Agency). The formation of NASA allowed Wernher von Braun to pry himself free of his military overseers and forge ahead with his lifetime goal of space exploration, but the Russian most identified as his counterpart, Sergei Korolev, would never entirely shed the yolk of Soviet military oversight. Korolev was forever besieged by competitive designers, engineers, scientists, soldiers, politicians, bureaucrats and worst of all dangerous intelligence mandarins who had literally threatened his life and made him wary of his Kremlin overseers.

In *Challenge to Apollo,* author Siddiqi makes a persuasive case for there being two different versions of Soviet space history. There is the version sanctioned and rubber stamped by the communist apparatus and then there is the version which came to light after the collapse of the Soviet Union. The latter version is far more colorful and complex and undoubtedly closer to the truth. Both the "official" history and the true history share many obvious similarities and they both certainly credit Korolev with a position of pre-eminence, although that was not the case during his lifetime. In fact Korolev would get no official recognition while he was alive.

In the last year of World War II, Korolev would be partially rehabilitated. His record was not yet expunged and so technically he was still a convict. The stories intimate that it was Glushko who informed the military of Korolev's importance, and suggested that he should be released and put to work. Even Korolev's mother had pleaded directly to Stalin to have her son exonerated, but although he had been moved from the Gulag in Siberia back to a less harrowing prison near Moscow, it would not be until August 1944 that he would finally be released, almost exactly six years after his arrest. Glushko would be released the same day. (Coincidentally, at almost exactly the same time, Von Braun was arrested by the Gestapo and held in prison for 12 days—also without charges.) In a clear indication of his passion for rockets, Korolev remained for another year at the same prison, working as a free man on rocket and jet assisted take off for airplanes (JATO). For a brief time he and Glushko worked together, with Korolev acting as Glushko's deputy. Then on September 8th 1945 Korolev was sent to Germany as a newly enrolled colonel in the Red Army. On his arrival he was integrated into an organization known as *Institut Rabe* (Rocket Manufacturing and Development), a secret operation whose task was to track down and capture German scientists and technology. The institute comprised over 1000 staff, half German, half Russian, with 50 people from Peenemünde and one key German POW, Helmut Grottrup. Just two days before Korolev's arrival, Valentin Glushko, who had been there since the Spring, had successfully fired his first captured V2 engine.

Korolev and his colleagues were astonished by Von Braun's A-4/V2 rocket, a 46 foot tall, 13 ton alcohol and oxygen propelled missile with a range of 225 miles. At the time, Russian rockets were, at best, generating a few hundred pounds of thrust, while the V2 was capable of 45,000 pounds. Despite his early work with Tsander, Korolev had shifted his enthusiasm to solid rockets—until he encountered the V2.

A month later Glushko and Korolev took a trip to Cuxhaven, a town on the North Sea about 100 km west of Hamburg, where they would be on hand to see a British launch of a captured V2 rocket. They reported what they witnessed back to Moscow and it swiftly became clear to Stalin and his sycophantic inner circle that this new rocket technology represented a formidable weapon. Stalin ordered the immediate scouring of the German countryside for every shred of equipment, paperwork or personnel, all with the intention of shipping them back to Russia. One thing in particular attracted the Soviet leader's attention, reports of a design for a Nazi space bomber capable of reaching New York.

V2 at Cuxhaven

At this very early stage a rivalry broke out between the Soviet Air Force and the Artillery Directorate. Both wanted to take control of this new missile technology and both thought they had valid claims to do so. The Soviet artillery had used the Katushya rocket system to great effect during the war and yet the Air Force had been using early rockets to help with primitive JATO. A similar rivalry had existed in the USA but it had all but vanished with the formation of NASA. The Soviet rivalry would continue well into the 1970's. The powerful artillery directorate would ultimately play the strongest hand and create the Soviet strategic missile infrastructure. The Air Force would take charge of the cosmonauts and rocket engine development. So in effect nothing changed, the Air Force would continue to build propulsion systems and train pilots to fly the vehicles, while the Artillery would continue to launch the rockets. It may have seemed an obvious solution at the time but it did not bode well for those who were interested in flying people on rockets.

The Artillery commanders had little comprehension of this advanced rocket technology and so they had to rely on men like Korolev and Glushko to make the technical decisions. This uneasy alliance would create a difficult organizational structure in the years ahead since it led to a sort-of cult of personalities, with each major player having his own design bureau, and each having his own benefactors within the military and the Kremlin.

An assortment of institutes and commissions were established over the next few years, all with the ultimate goal of understanding and duplicating the German A-4/V2. The largest of these new organizations was called *Institute Nordhausen* and Korolev was to be the second in command, a surprisingly elevated position for a recently released convict. Inside the labyrinth of bureaucracy at Institute Nordhausen was a veritable *Who's Who* of future Soviet space leaders. Glushko, Korolev, Vasiliy Mishin, Vladimir Barmin, Boris Chertok, Nikolai Pilyugin, Mikhail Ryazaniski and the German, Helmut Grottrup.

By the summer of 1946 Stalin had signed an order which would push missile development to the forefront of the Soviet military apparatus, this moment heralded the birth of the Cold War, predating the first Soviet atomic bomb by three years. The single most important figure in this early space bureaucracy was the Minister of Armaments, Dimitry Ustinov. It is tempting to find a counterpart in the west for Ustinov, but there really isn't one. Ustinov would wield unprecedented power over the future Soviet space program and there was no one in the west who had anything even approaching his level of influence. It is almost as though the NASA administrator, the head of the armed services committee and a member of the US Joint Chiefs of Staff were all rolled into one person. Such was Ustinov's role.

Despite assurances given to the contrary, in late 1946 the Germans captured at Peenemünde, including Grottrup, were forcibly shipped back to the Soviet Union and ensconced into one small aspect of a new organization which was called NII-88. This facility had branches that had been manufacturing armaments since the early part of the 20th century and was located near Kaliningrad, a suburb about 20 km north east of Moscow. NII-88 would ultimately evolve into the most famous facility in the future Soviet space program and would be variously known as OKB-1, TsKBEM, NPO Energia, Prime Design Bureau GKB and finally today as S.P. Korolev RSC Energia.

The Air Force, not to be outdone, took the remnants of RNII and established their own special design bureau for research into large rocket engines, this would be the beginnings of Glushko's empire later known variously as OKB-456 and *NPO Energomash*. It was located in a north west suburb of Moscow, called Khimki. Other members of the, now subsumed, Institute Rabe would find themselves in charge of different bureaus, each with a specific task. By the end of 1946 this group would be holding meetings and would ultimately become known as the Council of Chief Designers. Korolev in charge of missiles, Glushko for engines, Barmin for launch pads, Pilyugin for automatic flight controls, Ryazinsky for radio controls and Viktor Kuznetsov, who would take care of gyroscopes.

In early 1947 Korolev had already learned as much as he could from the German A4 and he was keen to push ahead with a much more advanced Russian design. The A4 assembly line was rebuilt and had been able to reconstruct 15 A4s from the remaining parts captured in Germany. After about a 50% success rate with the captured German A4s, word came from above that Korolev was to start construction on an all-Russian derivative, which was to be designated the R-1. This would be an almost identical copy of the A4, complete with a replica of the German engine, but this engine would be Russian, built by Glushko, and designated the R-100.

The creation of this first large booster involved over 35 different manufacturing plants and 13 research institutes. It would make its first successful flight on October 10th 1948, almost exactly one year after the Russians had launched their first German-built A4. Working with the benefit of advice from the captured Germans, the R-1 launch infrastructure was almost identical to that used during the war by the Nazis. The R-1 was transported and elevated on a special vehicle that used an erection armature to stand the rocket vertically. It was accompanied by an armored control car, hydrogen peroxide tankers for providing fuel to the turbines, peroxide warming units, liquid oxygen and alcohol tankers, transformer trailers, a fire fighting team and various electrical and battery trucks. The launches of both the captured A4s and the newly minted R-1s were conducted at a military facility called Kapustin Yar, halfway between the town of Volgograd and the Russian-Kazakhstan border.

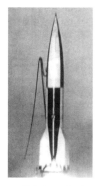

R-1 launch at
Kapustin Yar

The R-1 was launched twelve times with only three failures, its range was only slightly more than the A4 but it served to convince Ustinov and the other military leaders that Korolev and his team were making good progress. Meanwhile, the Germans were working up their own version of an advanced A4. This inevitably caused a problem as the Russians had their own plans for a

R-2 on pad at
Kapustin Yar

more powerful design of their own, called predictably, the R-2. A debate ensued about the merits of the two designs before it was decided to proceed with the Russian proposal.

While discussions continued, regarding the much more powerful R-2 launcher, Korolev and his team continued to fly and gradually upgrade the R-1. Variants included the R-1A which on April 21st 1949 became the first so-called geophysical rocket, as it successfully took scientific instruments to an altitude of 110 km and returned them via parachute. The R-1V took two dogs up to 101km and returned them safely on July 22nd 1951. This process continued throughout the summer and fall of 1951 with varying degrees of success and minor modifications to the payload portion of the rocket. Meanwhile, the R-2 made its first successful flight on October 26th 1950. The R-2 was about three meters taller than the R-1, although an experimental version flown a year earlier was slightly smaller. The main improvement over the R-1 was a form of primitive staging with the singular purpose of separating the warhead from the booster. Between October and December twelve R-2s were launched and all of them failed. However, later derivatives of the R-2 would be launched right up until 1960, most of them in pursuit of basic research.

The next booster on Korolev's table was the R-3 and it would incorporate some of the most fundamental changes yet. Korolev designed the R-3 to use its fuel tanks as part of the structure of the rocket, in much the same way as the Americans' Atlas missile. He also wanted to change the fuel from alcohol to kerosene, potentially adding another 20% to the rocket's lifting power.

Most western preconceptions of the Soviet system would preclude the idea of encouraging bidding for contracts, but that is exactly what Korolev did with the R-3, soliciting offers to supply the new, more powerful, engines required, not only from Glushko, but also from other smaller engine builders. The new engine would have to produce over 100 tons of thrust and the R-3 would stand almost twice the height of the R-1, with a proposed range of 3000 km. Enormous pressure was being exerted by the Kremlin to get an operational ICBM system deployed. It soon became clear that the huge advances necessary for the R-3 were too much to accomplish in the short time available and so the R-3 was shelved, in favor of a smaller derivative called the R-5. Much of the advanced work done on the R-3 by Glushko and Korolev would ultimately benefit later heavy-lift vehicles but in the meantime the incremental step represented by the R-5 seemed more prudent. Only marginally larger than the R-2, the R-5 would make its first flight in March of 1953. Gone was the

kerosene engine, replaced once again by alcohol, but
the booster was soon co-opted from its primary
purpose. Variants of the R-5 flew scientific payloads
under the designations R-5A, R-5B, R-5V and R-5R. In
1958 the R-5 achieved an altitude record and as late
as 1964 an R-5V was even used to test some equipment for the manned Soyuz program.

In early 1950, the promise of the R-3 had led another engineer, Mikhail Tikhonravov, to investigate the
possibility of using the launcher to send a payload to
orbital velocity. Tikhonravov had been speculating on
such things for nearly three years and when he
looked at the R-3 he realized that the rocket would
be powerful enough to make his notions feasible. He
mentioned these ideas to Korolev who embraced
them enthusiastically. The following month Korolev
persuaded Ustinov to allow him to restructure his
organization, changing his department from the NII-
88 Special Design Bureau Department 3 (SKB-3) to
Special Design Bureau #1(OKB-1), with Korolev in

R-5A on pad at
Kapustin Yar

charge as chief designer of missiles. By October of 1950 the Germans were
completely removed from all classified work. They were sent back to Germany
incrementally between January 1952 and 1954.

The Russian team now forged ahead independent of foreign influence, and by
October of 1951 Tikhonravov was even writing about trips to the moon in
the Soviet press. Just a month later the R-2 was officially deployed as a
weapons system and began to be manufactured in increasing numbers, despite
the fact that it required awkward fueling procedures. Problems with the R-3
delayed further progress and the decision to implement the R-5 model result-
ed in no new launch vehicles leaving the ground successfully until April 2nd
1953. The first two launch attempts of the R-5 failed but the third flight
reached its target. Over the course of 1953 fifteen R-5 launches would take
place, with only two failures. For the next two years Korolev and his team
began to further investigate the possibility of elaborate multi-stage boosters.
Meanwhile, Glushko was having considerable difficulty with his new engine, the
RD-110; the kerosene fuel generated considerably more thrust and also
equivalent increases in pressures. As far back as his days with the GDL
Glushko had found it difficult to work with super cold propellants such as liquid oxygen, and he was keen to drop them in favor of storable fuels.

As early as February 1953 the specifications had been nailed down for an even
bigger booster than the R-3. Its range would be 7000km, the thrust would be
over 270 tons and it would be an entirely original and distinct design, consist-
ing of a central rocket section with four strap-on boosters.

R-7 in flight at
Baikonur

In October of 1953 a report from a noted nuclear scientist would completely rewrite the specifications for the next generation of booster. It had to be able to lift the Soviet hydrogen bomb, which was considerably heavier than anything launched previously. The new booster was soon designated the R-7. Glushko's engines that were under development at the time (the RD-105 and 106) were not powerful enough to lift the necessary six ton payload and so a radical design breakthrough was necessary. This came in the form of an idea from one of Glushko's competitors, Alexei Isayev, a rocket designer at NII-88. Isayev realized that if the thrust were to be spread out amongst a cluster of combustion chambers, not only would one turbopump suffice but it would have a multitude of knock-on advantages, the thrust gain was considerably more than the weight increase of the hardware and the internal vibrations were minimized. This was arguably the most important breakthrough in Soviet rocket development and it came from neither Glushko nor Korolev but from a minor player, all but unknown in the West even today. With Isayev's modifications Glushko was able to construct the RD-107 and RD-108, the former providing 83 tons of thrust and the latter 75 tons. By combining four RD-107s in the strap-ons and a single RD-108 in the central booster the total lifting thrust of the R-7 package was in the region of 400 tons. There was nothing else anywhere on the drawings boards that was even remotely like it. Over the course of 1954 the booster's final design took shape while Ustinov worked the bureaucracy to move it to the top of the pile as a national priority. The overall appearance of the rocket, with the four strap-ons, was later attributed to Korolev's chief deputy Vasiliy Mishin.

Up to this time every Soviet booster had been a tactical weapon, which means it could be launched from just about anywhere, but it had a limited range. If the R-7 could be made to fly it would be a strategic weapon; its launching site could be just about anywhere but it could also reach almost any target on Earth. This introduced the concept of a permanent launch site, built far away from prying eyes. An almost farcical debate took place in which several choice locations were dismissed, including near to a very popular holiday resort, before a remote part of the steppe in Kazakhstan was chosen, near a huddle of buildings called *Tyuratam*. It was more than twice as far from Moscow as Kapustin Yar. It was (and still is) a very unpopular choice, mainly due to its extreme climate, with temperature fluctuations over any given year of up to 75 degrees centigrade. Nevertheless, it soon became the site of a massive and difficult construction project, given crucial support by the lone railway track that passed through the area. Over the next fifty years it would become the biggest spaceport in the world with an area of over 5000 square kilometers and would be most famously known as the *Baikonur Cosmodrome*. Even the name of this most famous launch site was clouded by ridiculous obfuscation. The original town of Baikonur was in fact nearly 300 km north east of the launch facility, but the name was chosen specifically to confuse western intelligence agencies. Adding to the confusion the town-sized facility which grew out of the project, originally known as *Site 10* was called Zarya, Leninsk and Zvezdograd before it was officially renamed *Baikonur* by Boris Yeltsin in the 1980's. However, the closest railway station continues to be known as Tyuratam.

Starting in August 1955 (over five years later than the establishment of America's primary launch site at Cape Canaveral) an army of workers gradually converted a desolate patch of Kazakh scrubland into a complex launch facility. That very same month, at the IAF congress in Copenhagen, both the USA and USSR announced they would try and launch a satellite by

R-7 launch pad at Baikonur

1957-58. Oddly, one of the most important meetings in the birth of the first artificial satellite didn't take place until *after* the IAF announcement. At that meeting Korolev was told that no thought would be given to a satellite until *after* he had proven the R-7's worth as a weapon delivery system.

During this whole period Tikhonravov continued to explore the difficulties of launching an orbital payload. He had switched his attention from multi-stage boosters to the design of the actual spacecraft. Keeping Korolev informed of his work, Tikhonravov was now thinking of satellites. In a stroke of bravado Korolev took a chance at selling the satellite program by pitching it directly to the top. On February 27th 1956 Soviet Premier Nikita Kruschev was to pay his first visit to Korolev's OKB-1 facility. Conveniently a model of Tikhonravov's prototype satellite was placed close to a full size mockup of the R-7. Taking his time, Korolev quietly explained to Kruschev that the R-7 could easily place a satellite into space by simply swapping out the payload. Kruschev took the bait and approved the idea as long as it didn't interfere with the development of the R-7 as a weapon. The specifications for the satellite had been defined just two days earlier.

Between March and October of 1956 the entire R-7 launch complex was assembled in a building in Leningrad before being dismantled and shipped by rail, to be reconstructed at Tyuratam. While all this was going on, the R-5 was continuing its string of successes out of the launch site at Kapustin Yar. The R-5M was officially deployed into the Soviet arsenal on June 21st 1956, just four months after it had successfully delivered its first nuclear weapon to a target range. Korolev and his team were using the R-5 to gradually refine guidance and other issues in preparation for the birth of the R-7.

As if Korolev didn't have enough to do, he had also spent the better part of the last three years perfecting another booster designated the R-11, which was the Soviet counterpart to the Polaris submarine-launched missile. The R-11 test program had proceeded relatively painlessly and it had been successfully fired for the first time from a submarine just three months before Korolev's encounter with Kruschev. All of this work was making it increasingly difficult for Korolev's organization to continue to operate efficiently with the distractions of NII-88 and so in August of 1956, Korolev's OKB-1 became a self managed and self contained organization answering directly to the government and the military. NII-88 passed into history.

The following month von Braun launched a Jupiter C rocket which could have—given the addition of an extra stage—launched a satellite into orbit. Korolev immediately assumed the Americans were about to upstage his plans for the R-7 so he redoubled his efforts. Adding to his problems, the prototype RD-107 engines were not performing at peak efficiency and so it occurred to him to rethink his strategy. Tikhonravov's satellite was large and heavy, clocking in at around a half a ton. Why not try something a little smaller?

Korolev's supporters for the satellite comprised a fragile alliance of artillery officers, politicians and scientists and when he summarily postponed the larger satellite, with its scientific payload, he inevitably met with some resistance. However, everything still hinged on the success of the R-7, which was yet to fly. At this moment in time the R-5 was capable of flying about 1200km and the Americans were secure in the knowledge that their Jupiter missile had flown almost five times further. This illusion of superiority would soon evaporate once the R-7 took flight.

Starting on May 15th 1957, the R-7 test flights began. The first booster failed and landed 300 km downrange, the second suffered several pad aborts and the third lost its strap-on boosters about a half a minute into the flight. However, on August 21st 1957 the R-7 established a new record for a ballistic missile, flying over 6500 km to its target on the Kamchatka peninsula. Korolev was delighted and immediately turned his attention back to the tiny satellite taking shape in his factory at OKB-1. Unfortunately there were still some detractors in the upper echelons of the military who were not at all interested in Korolev's desire to one-up the Americans with a satellite launch. Despite being ready he was obliged to fly the R-7 one more time on September 7th before getting the green light to try for orbit.

A special cone-shaped fairing had been designed to encompass the satellite, whixh was a small ball, just under two feet in diameter. By replacing the larger weapons' fairing the R-7 now took on a uniquely stunted appearance. The

modified R-7 booster was transported to the launch
pad on October 3rd. Inside the fairing the small met-
al sphere was mounted on a pneumatic separating
device that was designed to trigger about 20 seconds
after main engine shutdown. The whole stack
weighed about 267,000 kilograms with the satellite
making up less than half of one percent of that mass.
Around ten thirty in the evening, local time, on October 4th, Glushko's engines lit up the night with slightly less than the prescribed thrust of 400 tons, and the
R-7, carrying its payload designated the PS-1, left for
space. Just five and a half minutes later the transition-
al module fairing opened and the pneumatic piston
ejected the satellite into space with about 15 kilograms of force. Now Korolev and his team waited for
confirmation that they had created the world's first
artificial satellite.

The only way to know that they had succeeded
would be to detect the faint radio signal emitting
from the satellite as it passed overhead about ninety
minutes after launch. Tracking stations in the far east
had detected the signal almost immediately, but
Korolev would not be satisfied until he heard it
himself at Baikonur from a radio truck set up
specifically to detect the signal. The space age
began with the innocuous beeping of a small
radio transmitter flying in a highly elliptical
orbit. It was soon dubbed simply *Sputnik*, which
variously translates as "satellite" "companion"
or "traveler".

R-7 Sputnik launch
October 4th 1957

It had been only six weeks since the first successful launch of the R-7. The booster was
twenty times more powerful than the A4 and it
had evolved in just over a decade. It was capable of throwing nearly five tons around the
world but it would prove to be almost useless
as a quick response weapon. The propellants
took nearly five hours to load and once loaded
were extremely difficult to offload. In the very
near future, that amount of time in a war situation would constitute an eternity. Long before
the R-7 could be fueled and fired, its launch site
would be irradiated rubble. Although Korolev
didn't know it yet, time would swiftly overtake
his masterpiece. The only thing the R-7 was

Sputnik 1 in its fairing

R-7 Sputnik 2 on pad
at Baikonur

Sputnik 2 mockup
museum display

really any use for was space flight—and for that, it was supremely well-suited.

The flight of Sputnik was a major propaganda victory for the Soviet Union. Premier Kruschev was so delighted with the unanimously positive global response that he immediately ordered Korolev to repeat the trick in time for the impending 40th anniversary of the Russian revolution. By now most of the problems with preparing Tikhonravov's original satellite had been overcome and the thrust deficiency in Glushko's RD-107 engine had also been surmounted. Nonetheless, to meet Kruschev's deadline it was decided to forego the large satellite one more time and use a modification of the Sputnik 1 design, this time with the addition of a small compartment to hold a passenger. Korolev ordered the 508 kg satellite to be prepared for launch. This time Sputnik would have live cargo, a dog called *Laika*. To simplify the requirements for keeping the animal alive it was decided to keep Sputnik 2 (as it would become known) attached to the second stage of the R-7 booster once it reached orbit. On November 3rd 1957, not even a month after Sputnik, Laika was boosted into another highly elliptical orbit. The mass of the orbiting payload was a significant surprise to American experts, clocking in at over six tons. The dog survived for four days before excessive heat in the tiny cabin finally took its toll.

Even before the flight of Sputnik 2 the R-7 booster was undergoing the first in a series of important modifications. It was determined that to better improve the chances of launch success, and to eliminate potential problems, the main engines would be reprogrammed. At launch the central core of the rocket, which contained the RD-108 engine, would be throttled back to 80% at launch, while the four RD-107s would also be retasked to drop their thrust to 75% just before stage separation. This procedure of throttling back to alleviate dynamic pressure is still common practice on most space launches today.

Less than two weeks after Sputnik 2 burned up in the atmosphere, on April

27th 1958, this modified configuration was finally put to the test carrying Tikhonravov's large satellite which was known as *Object D*. Sadly, all the long work on the large satellite disintegrated just over a minute and a half into the flight, however, a fully operational back-up of Object D flew into orbit on May 15th 1958. Object D was swiftly renamed *Sputnik 3*. It weighed an unprecedented 1327 kg and carried an array of scientific instruments, including a magnetometer, two cosmic ray detectors and a mass spectrometer. Although Sputnik 3 lost contact after only 19 days it relayed tens of thousands of measurements back to the ground and convincingly proved that the Americans were not the only ones who could bring scientific results back from space. (*Explorer 1* had discovered the earth's radiation belts the previous February.)

Sputnik 3 mockup
museum display

Korolev's rocket soon revealed its true capabilities as the world's most practical and durable space launcher; but Glushko was not to be involved in the next upgrade. Korolev needed to add a third stage to the R-7 in order to reach escape velocity from the Earth. (In modern parlance the central core and the strap on boosters of the R-7 are considered the first two stages). An argument ensued between Korolev and Glushko over the choice of propellants that the upper stage should use. Glushko favored the highly toxic and problematic hydrazine while Korolev, always pressed for time, wanted to use the much safer and more familiar kerosene. In what turned out to be the beginnings of a major rift between the two chief designers, Korolev opted, in February 1958, to have the engine built at an independent bureau by Semyon Kosberg. By adding this small upper stage, dubbed the E-1, to the R-7, Korolev was able to begin humanity's first robotic sorties into deep space.

Three consecutive failures of the revised booster configuration, between September and December of 1958, did little to alleviate the animosity brewing between Korolev and Glushko, but on January 2nd 1959 the revised three stage rocket took the first spacecraft out of Earth orbit and on its way to the moon. The extra third stage used Kosberg's RD-0105 engine (which was the first Soviet engine with the ability to start in a vacuum) and was able to push the small 361kg sphere up to escape velocity. A minor malfunction of the booster served to interfere with the probe's primary objective, which was to crash into the moon. Instead the *Luna 1* probe became the first man-made object to be placed into solar orbit, where it made a series of important discoveries. It missed the moon on January 4th by just less than 6000 km.

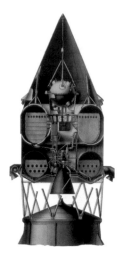

Luna 1 with E-1
upper stage

R-7/ Luna 1 launch
at Baikonur

On June 18th of 1959 the modified R-7 with the E-1 upper stage was ready to try for the moon again but once more failed just after launch. The initial flurry of success enjoyed at the end of 1957 was now beginning to look more like good luck, because most of the failures of 1958 and 1959 had little or nothing to do with the new upper stage. However, on September 12th *Luna 2* was finally dispatched to the surface of another world. This impressive achievement was swiftly followed up with the launch of Luna 3 on the second anniversary of the flight of Sputnik. Luna 3 was a much more complex craft than either of its predecessors. The outside of the satellite was enveloped in solar panels that supplied electrical power to the camera system on board. Unlike Luna 2, which had done little but plow into the moon carrying a bunch of plaques and flags, Luna 3 was carrying a sophisticated camera, scanning and fax system. On October 7th 1959, from an altitude of just over 6000 km, the Luna 3 photographed the far side of the moon. Its figure-eight orbit brought it back close to the Earth eleven days later at which point it transmitted the pictures to ground controllers. It was a remarkable achievement for the Soviet team, since the photographs showed the world the lunar far side for the first time in history. Despite the success of Korolev's advanced R-7 lunar program his decision to not use Glushko to build the upper stage engine would quickly fester into an all-out feud over the future direction of Soviet rocketry.

Less than two months after Korolev had given the contract for the E-1 engine to Glushko's competitor, Glushko had his own meeting with Premier Kruschev and convinced the Soviet leader that he should assign a major missile contract to one of Korolev's competitors. Undoubtedly under pressure from his chief military advisors to begin funding a more practical ICBM weapon system, Kruschev decided to go with Glushko's recommendation and in May 1958 gave the go-ahead to Mikhail Yangel's design bureau, OKB-586, for this new missile. Yangel had been the beneficiary of Korolev's earlier work with storable propellants and he would later put that research to good use. He had already contributed two strategic missiles, the R-12 and R-14 to the Soviet arsenal. The R-14 would go on to become the first stage of the *Kosmos* space launcher, used to place many smaller payloads into orbit, but the proposed R-16, with storable propellants and a greatly extended range would be an effective ICBM, and if Yangel and Glushko could make it fly, it could make the R-7 obsolete as a weapon. Korolev tried to convince the Soviet leadership that he could build a better ICBM, the R-9, but still using cryogenic liquid oxygen and kerosene. This competition ultimately placed two new boosters into the Soviet arsenal but more importantly Glushko's support for Yangel further increased the tension between the two most important players in the Soviet space program.

R-14, basis for the
Kosmos launcher

While struggling to prove the worth of his boosters to his military bosses, Korolev continued work on another upgrade to the R-7. This had begun life on the drawing board in January 1960 and was to be later named the *Molniya*. Taking the basic structure of the R-7, Korolev's team reinforced the vehicle to be able to withstand the additional structural loads of the vehicle. The R-7/Molniya would be a full eleven meters taller than the basic R-7 ICBM that had flown three years earlier. This additional height was comprised of two upper stages that were designated Unit I and Booster L. Unit I was basically just a third stage that was powered by a RD-0107 kerosene/LOX engine and built by the Kosberg bureau, the same bureau that had built the E-1 and RD-0105 (Note: engines beginning with RD-0 were Kosberg engines while those beginning RD-1 through 7 were almost all Glushko's). Booster L was powered by Korolev's own S1-5400A1 engine, ostensibly the first engine ever built with a closed-loop and powered predictably by LOX/kerosene. This engine could start in space and was thus ideally suited for long-range payloads such as interplanetary probes. The Molniya/R-7 would be Korolev's choice for both his Mars and Venus probes as well as for sending heavier payloads to the moon.

R-7/Molniya four stage launcher

Vladimir Chelomei

While Korolev was inaugurating the space age, another engineer, one whose name is still relatively unknown in the west, was test flying primitive cruise missiles out of Kapustin Yar. His name was Vladimir Chelomei and his contributions to Soviet space efforts would be comparable to Korolev's. Chelomei had suffered a series of setbacks in the mid 1950's as he was bounced from project to project, often finding his basic research given to other, less-deserving designers. Finally, in the summer of 1955 he was given a small factory in Moscow where he could begin to develop his ambitious plans. He wanted to send ion-propelled fleets to Mars, much the same as von Braun had proposed on Disney's Sunday night television shows. He also believed in another of von Braun's dreams, a winged reusable space shuttle. This winged spacecraft was a natural extension of Chelomei's work on winged cruise missiles. He had also been privy to the German designs of Eugen Sänger, whose *Silverbird* long range space bomber had so infatuated Stalin. Just as von Braun was showing off winged Martian landing vehicles, Chelomei was suggesting a space shuttle for the same purpose, he called it the *Kosmoplan*. There were of course military applications for such a manned shuttle and Chelomei was undoubtedly aware that the Americans were working on just such a vehicle, the *Dyna-Soar*, as the next logical step after their *X-15* program. Chelomei's military variant was called the *Raketoplan*. These ambitious designs were being touted directly to Kruschev just as Korolev was building his Molniya/R-7 in preparation for an assault on Mars and Venus.

After making an impassioned speech to Kruschev, Chelomei was inducted into the ICBM building business. Now the Soviets had three totally independent and competing bureaus building ICBMs, all with their eyes on space travel. While the United States government retained literally dozens of missile building contractors at this time, they were all operating on the free market principle where if you didn't deliver (for the most part) you didn't get paid. There was not supposed to be any such parallel in the socialist Soviet system and yet the bulk of these design bureaus were struggling to compete in an almost completely capitalistic way for considerably less funds than were available in the USA. Korolev had clearly worked miracles by bringing the R-7 to its current level of proficiency and Glushko had certainly contributed substantially to the large engines required. Then there was Mikhail Yangel whose design bureau had filled an important void in the need for quick-loading, quick-deploying ballistic missiles. The introduction of a third player to compete with both himself and Yangel must have seemed completely absurd to Korolev; in hindsight Chelomei was to later prove his worth, but only at great expense to Korolev.

At the same time that Korolev was creating the Molniya, he began a series of meetings with his primary assistants to discuss even more ambitious plans. Von Braun's Saturn program had been under development in the United States since 1957 and although it wouldn't fly for almost another two years, there was nothing like it being planned at OKB-1. Glushko advocated following von Braun's lead by strapping multiple ICBMs to one center stage. For this, he suggested seven of Korolev's R-9s for the first stage and four for the second stage. This configuration would have been three times more massive than von Braun's more modest Saturn I, which was an aggregate of Redstone and Jupiter missiles. Korolev had similar ideas and was leaning towards a vehicle of slightly larger proportions to that suggested by Glushko, around 1.5 million kg. This would have placed it somewhere between the Saturn I and the Saturn V in size. Over the course of the Spring of 1960 Korolev would sell this long-range plan for an ambitious space program directly to the Central Committee and by June they were ready to approve his ambitions. The new super booster program was now potentially up to 1.8 million kg and was slated to fulfill the Soviet needs well into the late 1960's.

Almost as soon as Korolev had secured support for his long-range plans he turned his attention back to the manned spacecraft that had been gradually brewing at OKB-1 since March of 1959 . It was designated *Object-K*, the "K" was an acronym for the Russian word for *ship*. In many respects Object-K was superior to the American *Mercury*. The cosmonaut was to sit in a dedicated descent module while the scientific and support equipment were placed in a separate instrument module. In the event of problems during assembly, this modular approach allowed for much easier access to critical systems. The US program would not adopt this modular method until the Gemini program. Consequently when something went wrong with Mercury it often required

Object-K (Vostok)

the removal of perfectly good hardware to access the problem. It also meant that everything had to be compressed into a much smaller volume, making the Mercury considerably more cramped than Object-K.

Since as early as 1955, Korolev and some of his team had been speculating on the possibility of building an advanced launch vehicle capable of putting a human subject into space. Now that the R-7 had been refined sufficiently to be able to carry an upper stage it was inevitable that the next course of action was to try and see if it could be modified even further to launch a heavier payload. Once again Korolev turned to the work of his old friend Mikhail Tikhonravov who had been spending some considerable "unofficial" time working the problem.

It should be noted that when Luna 3 was performing spectacular things above the moon, the Soviet government still had no official policy regarding space, or space exploration. Korolev's achievements up to this point had been useful propaganda stunts, but the Soviet infrastructure was still firmly in the hands of Ustinov and the Strategic Rocket Forces. The old artillery men saw little use for these extracurricular activities and were more concerned with perfecting long range missiles that could be launched quickly by using storable propellants. In January of 1960 Ustinov had bluntly reminded Korolev that his work on a spy satellite was a more important project than his manned spacecraft. Not to be deterred Korolev, rather craftily, decided to make them both from the same basic design.

At this time the United States was conducting most of its surveillance of Soviet missile efforts by using the ultra-secret high-flying U2 spy plane. On May Day 1960, just a few months after Korolev's triumphant lunar success, one of these U2 planes was shot down over Soviet territory, thus precipitating a major political crisis. One of the consequences of this crisis was the sudden urgency for surveillance methods that could fill the void left by the now vulnerable U2. Even though the United States' rockets were well behind the Soviets in lifting power, they were still able to pull the occasional coup. Just three months after losing Francis Gary Powers' U2 to a Soviet anti-aircraft missile, the United States deployed the *Corona* spy satellite system, thus moving the world of military intelligence gathering into a whole new realm; one that required space launchers. The enormous launch pads required for the R-7 were easy to see from orbit, this made them an easy target. The combined problems created by the United States consistently over-flying the Soviet Union's most secret facilities with the inability to easily hide an R-7 launch site started to work against Korolev, especially with his most illustrious benefactor, Premier Nikita Kruschev. Despite this, as time went by, the Soviets gradually realized the enormous advantage that the Americans had by flying spy satellites over Soviet territory. Space-capable boosters clearly had a military advantage and so Korolev's R-7 would continue to be useful.

Ever the tactician, Korolev designed a modified version of his proposed Object-K manned spacecraft and equipped it with cameras. It was a shrewd move by Korolev because it not only kept his R-7 booster in the game it also allowed him to continue to work on his manned ship, while simultaneously mollifying the military. The spy version of Object-K became known as *Zenit* but it would not fly successfully until April of 1962, by which time the Americans had been pulling in pictures from Corona for almost two years.

Zenit spy satellite assembly

Korolev's team finally rolled out a boilerplate version of Object-K, now renamed *Vostok*, and on May 15th 1960 it took flight atop an R-7, equipped with an E-1 third stage similar to that which had sent Luna 1 to the moon. The Vostok weighed just over four and a half thousand kilograms, more than double the large Object-D/Sputnik 3 of just two years earlier. It was equipped with an unusual solar array at the forward end as well as an ingenious system of thermal control shutters. The cabin was modifiable to include an assortment of payloads, including animals, and the instrument module was dominated by a small de-orbit engine, developed and built over the previous year by Korolev's old friend Alexei Isayev, the same rocket engineer who had "fixed" Glushko's RD-107 problems.

The first Vostok flight exhibited problems with the guidance and orientation system and when the de-orbit engine fired, the vehicle was facing entirely the wrong way; instead of returning to Earth it moved into a much higher orbit. Instead of a four day flight it remained in orbit for over five years. It was not an encouraging start. The next launch took place two months later and carried two dogs as passengers; this time the R-7 failed only nineteen seconds into the flight and both dogs perished. Only eighteen days later Korolev's team tried again, this fast turn-around time was indicative of just how streamlined the OKB-1 manufacturing process had become for making not only the R-7, but also the Vostok prototypes. This time the mission was much more successful and the two canine passengers were not only to complete eighteen orbits they would survive re-entry and the ejection of their life support compartment. Ironically, just a day earlier the Americans had successfully retrieved their first useful canister of film from *Corona*. These two flights vividly demonstrate just how close the race had become. Both countries retrieved their first orbital payloads within hours of each other, although the Western media didn't know that the Soviets had flown a living payload until some time later.

This success encouraged the Soviet team to schedule their first fully equipped and manned spaceflight attempt for the end of 1960. For the next two months the wild assortment of projects continued unabated at OKB-1. Just after the third anniversary of Sputnik, on October 10th 1960, Korolev tried to use the new four stage R-7/Molniya to send a probe to Mars. Kosberg's new third stage failed to place the payload in orbit but it is worth noting that this Mars launch was a full four years before Mariner 3, the first American attempt. Incredibly, another attempt was made with the Molniya/Mars package only four days later but it ended in the same way, with a loss during launch. In the grand scheme of things this was only a minor setback, but worse things were brewing.

The military chiefs of the artillery directorate were still pushing hard to deploy a viable ICBM, using storable propellants. Yangel's proposed R-16 had yet to fly, but only two weeks after the failure of Korolev's first Mars attempt the inaugural R-16 was sitting on the pad at Baikonur. Evidently, the day before launch it had been leaking fuel but there was a much more serious flaw inside

R-16 missile on pad

the second stage. On the night of October 24th 1960 an error in the command controls triggered the engines in the second stage causing the entire booster to erupt in a catastrophic explosion while the ground crews were still engaged in fixing the leak; 126 people were killed. The inquiry delayed the work of many people who were also involved with Vostok.

Naturally, this terrible failure of Yangel's ICBM was a grim reminder of the dangers of using toxic storable propellants that are highly volatile and ignite spontaneously on contact. Ironically, the R-16 would finally be deployed with dozens of them placed in silos and coffin launchers around the Soviet Union, but the main reason for its existence, the so-called storable aspect of its fuel and oxidizer would turn out to have little advantage. Once the nitric acid was placed in the rocket it could only be kept there for a relatively short time before the rocket had to be *rebuilt*. The loading and off-loading of propellants didn't have such an adverse effect on the R-7. One other side effect of the accident was that Chelomei increased his own influence as Yangel's star temporarily faded. The same month as the R-16 disaster Chelomei was given the Krunichev Machine Building Factory. This facility would become a force to be reckoned with in the years ahead.

Vostok would return to the launch pad on December 1st 1960, again carrying two dogs. This time there was a problem with Isayev's TDU-1 de-orbit engine and, although everything else worked just fine, the vehicle was deliberately destroyed to prevent the possibility of it landing on foreign territory. Only three weeks later (just three days after the first successful flight of Mercury-Redstone) the fifth Vostok was launched but failed to make orbit when Kosberg's third stage once again shut down early. This time the animals were saved by a launch escape system that had been suggested by Tikhonravov after the first Vostok failure. In an ominous prediction of things to come, the descent module had not separated from the instrument module. This problem would come back to haunt the Soviets many times in the years ahead. The two failures in December were making it look increasingly unlikely that Vostok was going to be safe to carry a human any time soon. Selection of a candidate cosmonaut continued until Yuri Alexeyevich Gagarin was moved to the top of the flight roster.

On February 4th a third attempt was made to launch the four stage Molniya/R-7, this time with a probe intended to investigate Venus. Once again one of the upper stages didn't fire and the probe was stranded in Earth orbit, however, only eight days later on February 12th 1961, Korolev finally managed to successfully launch his four stage Molniya/R-7. This time it would take the 643 kg *Venera I* probe to within 100,000 km of Venus. Instrumentation aboard consisted of a magnetometer, ion traps, meteorite detectors, cosmic ray detectors and radiation detectors. The whole craft was equipped with solar panels for power. The fourth stage engine worked perfectly, demonstrating space start capability and the use of ullage rockets for the first time. Although Venera-1 didn't make it into

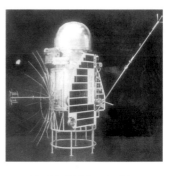

Venera 1 Venus probe

Venusian orbit it was another impressive triumph for the Soviets and all the more incredible when considered against the backdrop of such frenetic preparations for Vostok.

A final version of Vostok, one capable of carrying a human, was now complete. This model was designated Vostok 3A and it would make its first unmanned flight on March 9th 1961. The pressure to fly a successful mission was intense, since the Americans had flown *Ham* the chimpanzee into space just six weeks earlier. In keeping with an old tradition of not counting unsuccessful flights, this mission was called the *fourth* satellite-ship by the Soviet press even though it was the sixth. This spacecraft carried not only another dog but also a full-sized human mannequin. The mission lasted only one orbit, returning safely to Earth right on target. At this juncture history might have taken an entirely different course. Distressed by the problems experienced by *Ham* during Mercury Redstone 2, Wernher von Braun erred on the side of caution, choosing to launch at least one more unmanned vehicle on March 24th 1961—the flight was perfect. The very next day Korolev launched another dog and mannequin on a single orbit inside his second man-rated Vostok. The flight went exactly as planned and the stage was finally set for Korolev's greatest triumph.

On April 12th 1961 Yuri Gagarin climbed aboard the third man-rated Vostok spacecraft. It would become known as *Vostok I*, even though it was the eighth launch of the basic Vostok structure. A great deal of care had been taken in preparing Vostok I for Gagarin who had risen to the top of a promising group of Air Force pilot-candidates. At seven minutes past nine in the morning, Moscow time, Korolev's three stage R-7, carrying Vostok I, roared away from the launch pad at Tyuratam/Baikonur. Everything worked according to plan despite the checkered success rate of the three stage R-7. Just over eleven minutes after launch Gagarin became the first human to be placed in orbit. His mission profile was very similar to the earlier test flights, comprising just a single orbit. His 4725 kg spacecraft flew at an apogee of 327 km and a perigee of

R-7/Vostok launch

181 km; somewhat higher than predicted. After one hour and eighteen minutes the de-orbit engine fired and Vostok 1 began its descent. Unfortunately, in a replay of an earlier flight, the instrument module did not separate completely, but, fortunately for Gagarin, it never really endangered him and ten minutes later he began to reenter the atmosphere. At an altitude of 7 km the hatch blew and Gagarin ejected on schedule; he landed safely by parachute just one hour and fifty eight minutes after leaving Baikonur. The era of space travel had begun.

The sensation of Vostok 1's flight resounded around the world, but the Mercury astronauts training so feverishly in Florida were bitterly disappointed. Preparations were urgently made to counter this important Soviet victory and so, just over three weeks later, von Braun's team sent Alan Shepard into a 187 km high arc across the Atlantic. It was a historic moment, but compared to Korolev's achievement it seemed almost unimportant. Shepard's spacecraft weighed 1295 kg and barely scraped space for a couple of minutes. Gagarin's craft weighed three and a half times as much and flew around the world, only returning so quickly because it was programmed to do so. Although he didn't know it, this moment was the zenith of Korolev's career. Despite this amazing accomplishment, things would not get any easier for him as Glushko, Chelomei and Yangel prepared to deprive him of his space monopoly. Complicating the situation enormously was the proclamation which had been voiced by President John F. Kennedy just days after Shepard's flight. NASA was going to send men to the moon and they meant to do it in nine years. Korolev did not intend to settle for second place in that race.

Although the Vostok program seemed to be under firm control, the ambitious Molniya booster would remain problematic in its infancy. In fact out of the 26 times it flew between 1961 and 1965, about 50% failed totally while few of the remaining successful launches fulfilled their mission mandate. In early 1964 Molniya would be upgraded and would ultimately end up with an interchangeable third stage. This version would fly nearly 300 flights and would become known as the *Molniya-M* booster.

On August 6th 1961 another R-7/E-1 booster combination took Vostok 2, carrying the second Soviet cosmonaut, Gherman Titov, into orbit. This time the cosmonaut was to stay in space for one day, one hour and eighteen minutes. Once again during the final leg of the mission the instrument module failed to separate from the cabin. Despite this problem it was not quite as serious as that faced by Titov's counterpart in the West. Just a month earlier America's second astronaut, Gus Grissom, had flown a similar mission profile to Alan Shepard's. Shortly after splashdown the hatch on *Liberty Bell 7* blew prematurely and the capsule sank to the bottom of the Atlantic, almost taking Grissom with it. By comparison Titov, bailed out, parachuted down to the village of Krasniy Kut and immediately indulged in a bottle of beer!

As 1961 drew to a close there was a period of relative calm in Korolev's launch schedule. The next manned flight was a year away and so was the next attempt at reaching the planets. Just five days prior to Titov's flight the Soviet Central Committee had renewed their interest in Chelomei's ideas for two ICBM/space launchers that used storable propellants. He called them the UR-200 and the UR-500. In November Chelomei approached Glushko to provide him with engines for his proposed big boosters. Glushko apparently welcomed the opportunity to assist Korolev's competitor.

The following February Chelomei sold the idea for the UR-500 directly to Kruschev, who subsequently approved its development. Both the UR-200 and the UR-500 were to be built at Chelomei's Krunichev factory. Chelomei seemed to have outmaneuvered Korolev by simply promising the Soviet leadership what they wanted; a booster that could be an effective ICBM as well as a valid space launcher. Funding had literally been pulled from Korolev and transferred to Chelomei, which in hindsight seems quite incredible since Chelomei had yet to build any kind of booster on the scale of the R-7, much less fly one.

Chelomei soon returned to his ambitious plans for a spaceplane which could be used, not only for manned Mars missions, but also as an answer to the American X-20/Dyna-Soar orbital bomber. He had been working quietly since 1960 drawing up the specifications for his winged spacecraft, and by the end of 1961, when Korolev was regrouping, Chelomei launched his prototype aboard one of Yangel's R-12 missiles from Kapustin Yar. The 1750 kg MP-1 spaceplane successfully traveled a suborbital arc on December 27th 1961 to an altitude of 405 km before reentering at

UR-500

almost 4 km per second and landing intact near Lake Balkhash. Astonishingly the vehicle was completely intact and showed only minor heat damage. The prototype spacecraft had been conceived, built, and flown, in almost total obscurity, all in the space of eighteen months.

All through the Spring of 1962 the Americans made important strides toward catching up with the Russians. In February, and again in May, the Mercury finally made it into orbit atop the hastily retooled Atlas ICBM. Neither John Glenn's nor Scott Carpenter's flights were without mishap. Glenn had to reenter with his retro-pack attached, an eerily similar problem to that faced by Gagarin and Titov, while Carpenter reentered with almost no maneuvering fuel and was believed dead by the American public for several long hours before he was recovered hundreds of miles away from his intended landing site.

Korolev was painfully aware that the Americans were developing high powered cryogenic engines for their ICBMs. In fact the use of liquid hydrogen as a propellant was now finally becoming feasible. It had been known since the end of the 19th century that liquefied hydrogen (LH_2), burned in conjunction with liquid oxygen, would provide the biggest "bang for the buck" because of the potential energy released in relation to the weight of the propellant. Up until this time it had not been believed possible to work with LH_2 but an American contractor, *Convair*, who had built the Atlas missile, had been working painstakingly toward perfecting a hydrogen engine since 1956. In May of 1962 a first attempt had been made at flying this engine on the *Centaur* upper stage but the test was a failure. Korolev also knew that if he was to get the sort of thrust necessary to launch a manned lunar mission he was going to need the most efficient engines possible. However, he was fighting a losing battle because engines that used liquid oxygen with kerosene or alcohol were under siege, since they couldn't easily be used to quickly launch an ICBM. A hydrogen engine would do nothing but complicate and delay an ICBM even further since the handling and utilization of the super-cooled liquid was considered to be much too difficult.

Korolev pleaded with Glushko to build him a high-power cryogenic engine for his proposed lunar launcher but Glushko was adamant that he was going to stick with the toxic mixtures of nitric acid and hydrazine. Korolev had no choice but to look elsewhere. However, Glushko did not want to be shut out of any large scale rocket program and so he went over Korolev's head to try and force the use of storable propellants. Since Glushko had already committed to building a large engine for Chelomei's giant UR-500 he hoped to force Korolev to accept his position so he could use the same engine for both. The day after John Glenn's flight, Korolev and Glushko finally argued themselves into separate corners. Korolev decided to do the job without Glushko and so he turned over the engine construction to a less experienced design bureau.

All through the Spring of 1962 Korolev and his OKB-1 team worked on refining the details of their new super booster, which was now designated the N-1. The scale of the project would totally dwarf the R-7. Unlike in America where there seemed to be a gradual evolution from smaller to larger rockets culminating with the Saturn V, the Soviets were aiming to bypass the equivalents of the Titan and Saturn IB (both of which were used for manned flights) and jump straight to a 360 foot tall behemoth. This bold decision prompted some concern amongst Korolev's peers and so smaller variants of the N-1 were proposed, but they would never fly. By early summer Korolev was forced to defend his desire to use high energy cryogenic propellants and Glushko was there again, trying to convince the Committee otherwise. This time the cards fell in Korolev's favor causing the rift between the two designers to grow even wider. The N-1 booster program would be the largest ever undertaken in the Soviet Union and it would be mainly contracted to all of the same designers that had worked on the R-7, with one obvious and notable exception. Glushko would be replaced by Nikolai Kuznetsov's OKB-276 bureau, which would supply the engines.

That very same summer Tikhonravov's plans for an "upgraded" Vostok came back onto the table. It had become clear that something much bigger than the old Vostok would be necessary to send humans to the Moon or Mars and so a multi-module successor had been discussed. Korolev wrestled back and forth between an array of possible configurations that involved multiple dockings, earth orbit rendezvous, lunar orbit rendezvous and other methods. It was assumed that whatever vehicle was chosen it would likely have to dock in earth orbit and refuel, a procedure that was also causing sleepless nights in the von Braun household. These talks ultimately gave birth to the prototype of Russia's most successful spacecraft, the Soyuz.

While these issues were being debated the Vostok and planetary programs blossomed back to life. On August 11th the third manned Vostok raced into the sky above Baikonur to be followed just under 24 hours later by Vostok 4. Andrei Nikolayev and Pavel Popovitch spent just under four and three days, respectively, in orbit. Their joint flight seemed a brazen publicity stunt designed to remove some of the luster from the much more visible Mercury program. It did, however, have much more practical goals, Korolev now knew he could fly two ships simultaneously and that they could, with considerable care, be brought within close range of one another. This would be critical if Soyuz was to fulfill its mandate and a Russian was to walk on the moon.

The flush of success from Vostok was soon quashed when just two weeks later, on August 25th 1962, another Venera/Molniya failed to reach Earth orbit, amazingly this failure was repeated again on September 1st and again on September 12th. By October 24th 1962 a third Mars probe exploded in orbit before one finally made it to Mars distance, with a launch on November 1st 1962. This stunningly poor performance by the Molniya must have been a huge burden to Korolev. Just three days after getting Mars-1 off the ground yet another Molniya failed, destroying a second Mars probe in less than eleven days. As the new year began Korolev's team switched their attention back to the Moon. On January 4th 1963 a Luna probe ended up stuck in Earth orbit and on February 2nd one ended up spattering the Pacific ocean with debris. Finally, on April 4th, a third attempt made it out of orbit but missed the moon by over 8500 km.

While this difficult string of failures continued, a group of Soviet politicians were concocting another publicity coup. It was decided that a woman should be launched into space at the earliest possible date. Considerable resistance to this idea emanated from the design bureaus and the Air Force, but the idea was sanctioned from lofty heights inside the Kremlin and so the next Vostok mission would involve another double flight, but this time one of the pilots would be chosen from a select group of highly trained female candidates.

On the 14th of June 1963, just one month after the last Mercury astronaut, Gordon Cooper, had flown 22 orbits in his *Faith 7* Mercury capsule, Valeri Bykovsky departed for what would be a five day flight. Two days later Korolev's team notched up another first in the record books when they successfully placed Vostok 6 carrying 26 year-old Valentina Vladimirovna Tereshkova into orbit around the Earth. The flight did little to increase the Soviet's understanding of space flight in general, but it did score another major political victory for Korolev and his team. It was a triumphant end to the Vostok program. Despite the long string of failures that the R-7/Molniya had suffered in 1962-63, no one had been injured by an R-7/Vostok. When Tereshkova parachuted to safety no one suspected that the Vostok program would end with her flight. There were in fact extensive plans to fly several more Vostoks to higher altitudes, some with research animals. These plans were to "fill in the gap" while the next generation of spacecraft was being prepared. However, a situation developed which was very similar to that which had arisen in the USA.

Mercury had been a good beginning but the small capsule was barely capable of doing much more than flying around in circles. If anyone was to actually get to the moon it would require learning an entirely new skill set, including rendezvous, docking, space walks, ground tracking of multiple vehicles, celestial navigation procedures, and of course actually building the hardware. They would need fuel cells, spacesuits, cold-start engines even new kinds of food. It was abundantly clear that Apollo would not be ready for several years so the task was handed to NASA engineer Jim Chamberlin and his team to make an upgraded Mercury spacecraft, to be called Gemini. It was to be capable of enabling and developing all of these new technologies. In the Soviet Union, Korolev's lunar vehicle was even further behind than Apollo and he faced the same problems as Chamberlin.

Korolev didn't have the budget or the support to instigate an entirely new program like Gemini, and Gemini was to be launched on an ICBM using storable propellants, an option that Korolev wouldn't even consider. His rival Chelomei even started designing a spacecraft that looked a little like a cross between a Gemini and an Apollo, to be part of a military space station, and Chelomei was firmly in the dual-purpose ICBM/space launcher camp. It was rumored that pressure began to be exerted from the Kremlin to counter the Americans' plans by launching a manned flight with, not two, but three occupants. It is still the subject of some debate as to whether Korolev actually wanted this development to take place. He was already inundated with projects, primarily for the military, and he desperately wanted to concentrate on his lunar booster and spacecraft, but there are some who say he couldn't stand to be upstaged by the Americans and so he embraced the plans to create a multi-man spacecraft from the shells of the remaining Vostoks.

On November 11th 1963 OKB-1 lost another Molniya/Venera probe in Earth orbit. Apart from military work this concluded Korolev's space launches for 1963. A minor triumph occurred on 30th January 1964 when the three stage R-7 managed to launch two Elektron science satellites simultaneously from one launcher. The two small satellites were designed to examine the Earth's radiation belts in preparation for long duration manned missions. On February 4th 1964 Korolev was formally instructed to convert the Vostok for advanced uses. Between February 19th and March 27th a further four Molniya vehicles failed to deliver their four Venus-bound payloads. At this point any hope of making the R-7 a viable interplanetary launcher must have been seriously in question. Just six days later another Venera managed to get off the ground and on its way to Venus, this time it was renamed Zond-1. It was originally intended to deploy a lander onto Venus but the vehicle lost contact just over a month into the flight. Two weeks later, yet another unmanned Russian lunar lander failed to reach earth orbit. Meanwhile, the Americans had successfully launched the first unmanned Gemini.

The consistent failure of the Molniya four stage R-7 was somewhat offset by the successful launches of a more powerful three stage R-7 that later became known as the Voskhod launcher. The R-7 would be equipped with a third stage (I module) powered by the reliable and more potent RD-108 engine, instead of the usual R-0109. This configuration was launched successfully for the first time in late 1963 carrying Zenit-4, a military spy satellite that Korolev had built around a Vostok shell. The next three flights of this R-7/Zenit booster were on May 18th 1964, July 1st and September 13th. All carried similar Zenit satellites and all worked perfectly.

The first flight of an Apollo boilerplate atop a Saturn I booster on May 28th 1964 seemed to demand a response from the Soviets. Countering Kennedy's lunar challenge would be a formidable undertaking, with no guarantees of political coups as a side-benefit. Making it even more difficult, Korolev would find himself trying to assemble his giant N-1 lunar rocket without the benefit of Glushko's expertise, while simultaneously staving off Yangel and Chelomei's ambitious counter proposals. Chelomei had tried to continue work on his space plane concept, to fly shuttles to the moon and back, but the winged technology of the time simply couldn't hold up to the stresses of a full lunar-return re-entry. Clocking in at around eleven km per second the heat stresses were problematic even for an ablative vehicle like Apollo. It was at this time that Chelomei decided to shift gears and change to a ballistic re-entry vehicle.

Korolev was given a fully funded green light for the N-1 lunar booster within weeks of the first Saturn/Apollo launch. Then almost immediately the plug was pulled on his plans for a circum-lunar mission, instead funding was given to Chelomei to pursue his version of a lunar orbital mission which was to be called the LK-1.

Korolev had been working towards an ambitious lunar landing for years. He was expecting to use the same method as von Braun, using many launches to build a huge vehicle in earth orbit before deploying a very large lander on the lunar surface. Lunar bases would follow. Just as von Braun had to give up on these grandiose plans, so did Korolev. Unlike the R-7 his N-1 was going to be simply too big and expensive to launch several times per month. He had even spent several years gearing the N-1 towards an industrious Mars program, again, exactly like von Braun would do with his Saturn V. But whereas von Braun was unchallenged in his domain, Korolev soon found himself running side-by-side with Chelomei who had been given the contract for the lunar-orbiting mission while Korolev was to retain the landing mission. An analogy that doesn't even come close to the absurdity of this would be as if Boeing were told to build and fly Apollo 8 while Lockheed would build and fly Apollo 11; neither would use the same equipment.

Before any of these wildly ambitious plans could be realized there was the problem of keeping people flying. The 100% success record of his upgraded R-7/Zenit booster during the summer of 1964 must have encouraged Korolev to employ this configuration for his most ambitious manned flight yet.

The first step had been to gut the Vostok and redesign it to accommodate three passengers. The ejection seat was removed and replaced with three couches. There was no time to completely redesign the instrumentation so the crew would be obliged to look over their shoulder to see the flight controls. Also there would be absolutely no room for space suits or ejection seats. On October 6th 1964 Vladimir Komarov, Konstantin Feoktistov and Boris Yegorov raced to an apogee of 336 km aboard the singularly risky *Voskhod* spacecraft. The Russians were flying three men at once while America's two seater Gemini was still awaiting its second unmanned launch. However, the Americans had one or two aces up their sleeve and before the end of 1964 they managed to get their second major planetary probe into the air. Mariner 2 had successfully flown past Venus in late 1962 and now on November 28th 1964 Mariner 4 was on its way to making one of the greatest planetary voyages in history. It was destined to rendezvous with Mars the following summer.

Meanwhile, Korolev launched another unmanned Voskhod on February 22nd 1965. This time things didn't go quite as well when two simultaneous but separate commands from the ground *accidentally* set off the ship's self-destruct mechanism. It was the kind of error that could easily cause nightmares for the designers. Thankfully no one was aboard, but the accident precipitated some difficult decisions to scale back the very risky Voskhod program—but not before one last flight.

The Voskhod that had exploded had been testing an entirely new hatch and airlock mechanism. The hatch was built directly above the left couch and it was designed to allow a cosmonaut to step out into an airlock. This revised version of the Voskhod spacecraft had a 100% failure rate when Colonel Pavel Belyayev and Lt. Colonel Aleksey Leonov climbed aboard. Just over three weeks after the catastrophic failure of its sister ship, the three stage R-7 took flight, carrying Voskhod 2. It was March 18th 1965 and Leonov had a date with history. Korolev knew that a lot of his political power resided in his ability to beat the Americans and consistently stage publicity coups for the Soviet leadership. This may have played a part in the hurried timing of Voskhod 2. It was well known in the American press that the Gemini was being prepared to finally fly with a crew. There had been public discussions about the vehicle's ability to open its hatches while in space, leading to the inevitable talk about when the first space walk might take place. As it turned out, that first space walk was to happen hundreds of kilometers above the Caucasus, and Gemini would have nothing to do with it.

The Voskhod airlock was a remarkable piece of engineering that had been designed and built in record time. It resembled a large concertina of fabric that extended away from the new external hatch. Almost as soon as they reached orbit the airlock was deployed and ninety-two minutes after lift-off Leonov pushed his head out through the open outer hatch of the air lock and became the first human satellite. There have been many dramatic reports written about the difficulties encountered by Leonov when he became the first human to walk in space. After a dozen minutes of free flight he encountered problems re-entering the outer hatch. The suit he was wearing had inflated because of the large pressure differential and was now almost too large to fit through the 70 cm wide opening. He had to climb in head-first which presented an entirely new problem; he now had to rotate himself in the confines of the tun-

nel-like airlock. By all accounts his heart rate and temperature were at dangerous levels when the hatch finally closed and the airlock repressurized. Leonov climbed back through the inner hatch but after returning to his seat it seems the inner hatch exhibited a minor leak. After it was determined that this leak would eventually become an unacceptable hazard the decision was made to de-orbit the vehicle. This procedure was to have all been automatic but when the time came, the engine didn't fire.

The decision to cram several people into a Vostok would now have repercussions. Because the couches were oriented at 90 degrees from the way that the craft had originally been built, to be able to perform the necessary maneuvers manually, it would mean that Belyayev would have to lie horizontally while Leonov scrambled underneath

Voskhod 2 with airlock projecting at left

the couches holding his copilot in place. This was especially difficult now that they couldn't remove their space suits. Lying at right angles across the couches Belyayev was now looking forward, out of the same porthole that Gagarin's couch would have been facing on Vostok 1. Belyayev made the adjustments using the Earth as a reference point. The two men then scrambled to get into their seats before firing the engine. Thankfully Belyayev had done his job well and, although the time to reseat themselves caused an overshoot, they both survived. In a final ignominious ending to a historic flight the two cosmonauts found themselves camping in a snow covered forest for almost 48 hours while rescue teams scrambled to reach them. Even though five more Voskhods were approved for construction, they would never fly with people aboard. More significantly it was the end of an era for Soviet manned spaceflight. From this point onward Russia's cosmonauts would only fly in the Soyuz.

While Korolev was suffering near apoplexy at the antics of Leonov and Belyayev, his rival Vladimir Chelomei was continuing with work on his large UR-500 booster and its companion LK-1 circum-lunar spacecraft. At OKB-586, Mikhail Yangel was building ICBMs but was also showing an interest in

building a large space-faring vehicle. Yangel had been Korolev's deputy and later had been given his own bureau. Incredible as it seems, there were now three competing plans for a Soviet lunar mission. Even the inscrutable power corridors of Washington hadn't concocted anything so bizarre.

Five days after Voskhod 2 returned to Earth the Americans stepped up the pace and launched Gemini 3 with John Young and Virgil "Gus" Grissom aboard. The parallels between Vostok and Mercury are obvious, they both comprised six flights of single seat vehicles with little or no steering capability. Even the

Mikhail Yangel

parallels that would develop between Soyuz and Apollo are self evident, but there is simply no way to compare Voskhod and Gemini. While the Soviets had wasted time modifying a patently unsuitable spacecraft to stage two publicity stunts, the United States had methodically assembled a spacecraft that handled like a race car. Even Korolev's right-hand man, Vasiliy Mishin, admitted years later that Voskhod had contributed nothing to space research. With one notable and tragic exception the Soviet manned space program would now grind to a halt for almost three and a half years while the Americans would blaze ahead, flying nine more manned Gemini missions in seventeen months.

Korolev's original vision for an *orbital* lunar craft had been offered as an alternative to Chelomei's LK-1. It consisted of three stages and was rather confusingly referred to as the L1. Then there was the L2 configuration which was an unmanned lunar roving vehicle. L3 included a modified crew compartment with an attached lander, this version required four launches to accomplish. An

L4 mission was just a manned lunar orbiter and the final type, L5, would carry an advanced lunar rover. These motifs were quite different to the American approach which was a linear progression, the Apollo mission types were designated A – J, with each one improving on its predecessor, culminating with the J missions. The Russian L1 through L5 motifs were to be applied as and when necessary. So although L1 was an orbital mission and L3 was the manned landing mission, the other three motifs were supplemental flights that could be used as re-supply missions or simply as stand-alone delivery of hardware.

The key to Korolev's extensive lunar architecture was the 7K spacecraft which later became known simply as the *Soyuz*. Korolev fully expected that an L1 or L2 mission would require six launches by an upgraded R-7 while the L3 landing would require an upgraded R-7 and as many as three N-1s. The reason for such a hefty launch schedule was because he was originally shooting for an Earth-orbit rendezvous profile, with a very large 50-ton lander. Korolev's L1 orbital motif was in direct competition with Chelomei's LK-1/UR500 configuration and so it was cancelled almost as soon as Korolev's L3 landing mission had been given the green light.

The end of the Voskhod flights also heralded the end of Kruschev's domination of Soviet politics. Korolev and Kruschev had been at odds in recent years and so when he fell from power it was less of a blow to Korolev than it was to Chelomei who had been deriving some benefit from having Kruschev's son working at his bureau. Chelomei soon found himself being methodically stripped of his assets, including one important division originally known as OKB-301 which became today's *Lavochkin* agency, and would take over the Venera, Luna and Mars planetary probes. Meanwhile, Chelomei's large UR-500 somehow managed to survive the attacks coming from the Kremlin. In fact at the end of 1964 he was able to devise a larger version which he called the UR-500K that included a third stage and would be able to pull off an Apollo 8 style mission without any need for supplementary launches. The LK-1 spacecraft chosen for this mission had a similar shape to an Apollo CSM but with large solar panels attached. Superficially, at least, this is very like one of the configurations being examined today, but Chelomei's craft was only a single seater, while the proposed NASA CEV is expected to carry up to six people.

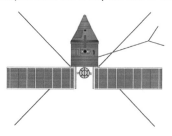

LK-1 proposed lunar orbiter
© Courtesy Mark Wade

The Baikonur cosmodrome was a massive facility spread over a vast tract of flat land in Kazakhstan. Heading northwards from the (new) town of Baikonur, the first thing a visitor encounters are tracking stations and radio masts. Continuing north the road sends branches east and west. The west road heads towards the giant UR-500 complex built specifically for Chelomei's boosters in 1962. Taking the east road you come to the Vostok launch site. Back on the main northern road from Baikonur the huge assembly buildings for Soyuz and the N-1 super booster are conspicuous, and beyond them lies the Soyuz and N-1 launch pads. The first UR-500 booster was successfully launched from the Baikonur complex on July 16th 1965 carrying a satellite called *Proton 1*. The UR-500 performed perfectly and marked Chelomei's first orbital launch—using storable propellants. The UR-500 would soon adopt the name of its first payload and become known as the *Proton*.

Korolev and his deputies spent much of 1965 and 1966 whittling away at their Apollo-class spacecraft. Since the circum-lunar mission had now been given to Chelomei, Korolev's lunar spacecraft continued on as Soyuz. It was to have three principal modules instead of Vostok's two. The stack consisted of a living module on top, the crew cabin and re-entry vehicle in the center and the service/instrument module at the base. Soyuz would weigh in at about six and a half thousand kilograms and would stand almost eight meters tall. Somewhat ironically the de-orbit engine for Soyuz would be powered by storable propellants. The launch vehicle would again be the ubiquitous R-7, but with a slightly more powerful "I" third stage that used the RD-110 engine instead of the RD-108. Soyuz would also have a launch escape tower, the first time any Soviet manned vehicle had such a basic safety device. This new R-7 variant is known today by the same name as its primary payload, *Soyuz*. The Soyuz stack would stand a full 20 meters taller than the R-7 that had launched Sputnik, but the first two stages remained essentially the same, a testament to the capability of the ten-year-old design.

One of the major problems with the Soyuz, faced by Korolev and his team, was the hazardous landing back on Earth. Gagarin and his five Vostok compatriots had been obliged to eject at around 7000 meters because the capsule faced a potentially hard landing, despite the parachute system. So with Soyuz an advanced landing system was created that would use a special altimeter to fire four, small, solid rockets an instant before landing, thus considerably softening the final impact.

Soyuz 7K

Meanwhile, Korolev now had to figure out how to get his L3 lunar landing motif into space. The reduction in funds to OKB-1 had forced a dramatic rethink in the architecture, because the N-1 simply wasn't big enough to carry the LOK orbiter and a lunar lander (the LK) to the moon as well as a crew of three people. It was decided to solve the problem from both ends.

Make the booster more powerful and more efficient *and* make the payload smaller and lighter (and effectively less safe.) One of the most obvious things that came out of this redesign was a similar solution that von Braun had used. When von Braun was faced with the ever increasing mass of the Apollo spacecraft complex he apparently made the decision to add another F-1 engine to the center of the Saturn V first stage. Korolev did the same thing but in the case of the N-1 the first stage went from 24 to 30 engines. It was increasingly apparent that Glushko's continued resistance to building large cryogenic engines was going to haunt the Soviet program for years to come. In fact, just before the end of Kruschev's reign, Glushko and Chelomei had pitched an engine, spacecraft and booster system that was specifically aimed at taking the

N-1 on pad

whole lunar program away from Korolev. Once again funds were sapped from OKB-1. Meanwhile, the only way that Korolev had managed to get support for his N-1 was by using it to derive new military applications. One of these was to have been the test-bed for the N-1 engine program but even that was cancelled. Korolev had been refused the funds to build a test stand big enough to house the first stage, and so the engines were never tested all-up until the first time they flew. This ridiculous state of affairs was in sharp contrast to the extensive F-1 engine testing that was taking place every couple of weeks in Alabama and Mississippi.

The lack of high power cryogenic engines forced the N-1 to start looking more and more like Jules Verne's space-train. Without the additional power of liquid hydrogen, the OKB-1 team was obliged to rely on more kerosene engines and more stages. Evolving over six years of designs, the final N-1 with its revised L3 lunar landing payload consisted of five stages below the spacecraft complex. The stages were configured as follows:

First stage: Blok A	30 m tall 16.8 m diam.	30 x NK-15 engines	4615 tons thrust
Second stage: Blok B	20 m tall 10.3 m diam.	8 x NK-15 engines	1432 tons thrust
Third stage: Blok V	11.5 m tall 7.6 m diam.	4 x NK-21 engines	164 tons thrust
Fourth stage: Blok G	8 m tall 4.1 m diam.	1 x NK-19 engine	40.8 tons thrust
Fifth stage: Blok D	5.7 m tall 3.7 m diam.	1 x 11D58 engine	8.5 tons thrust

The L3 lunar vehicle was then comprised of three more components:

Lander (LK)/Blok E	5.2 m tall	1 x RD858 & 1 x RD859	2 tons thrust each
	2.2-4.5 m dia.		
Orbiter(LOK)/Blok I	10 m tall	1 x 5D51 engine	3.4 tons thrust
	2.9 diam.		
Escape tower:	6 m tall	2 solid fuel engines	

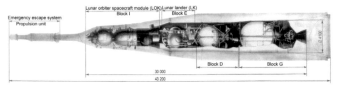

L3 lunar landing aggregate with LOK orbiter, LK lander and upper stages

The Saturn V by comparison had only three main stages capped off by the Apollo spacecraft and a launch escape system.

While the LOK orbiter would be the antecedent of the Soyuz spacecraft the LK lander was comprised of a landing stage, a crew cabin and an engine module designated Blok E, all to be built by Yangel; including the engines. The Blok E engine was used for both landing and lifting off from the moon. It had two engines that both fired simultaneously but once the computer determined the primary engine was working, the secondary engine would shut down. The Blok D stage would be used for velocity changes in lunar orbit and would take the LK lander down to within a few kilometers of the surface before dispatching the lander for final approach, so in that respect it was performing the function covered by the Apollo SM engine and the LM descent engine. On Apollo the lunar module had two separate engines; one for landing and one for take off, while the descent stage and its engine were left behind on the moon. On the LK the same engine was used for both take off and landing. It fired through a hole in the descent stage and then returned to orbit with the ascent stage, leaving just a donut with legs on the lunar surface. Once the LK returned to lunar

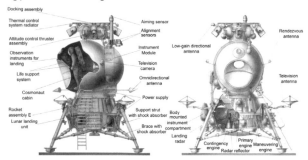

LK lunar lander

orbit it docked with the LOK but there would be no transfer tunnel so the cosmonaut would have to EVA to get back to the orbiter. The lander could only sustain one cosmonaut and at one point it was discussed that there might be two separate landers, one would take the cosmonaut down and the other would bring him home. He would travel from one to the other using a robot rover.

By August of 1965 the Soviet leaders were beginning to realize the waste of running two competing lunar programs. Korolev led an attack against Chelomei's lunar spacecraft, which had been mired in design problems for over a year. The result of this was that Chelomei's UR-500 would be combined with Korolev's LOK lunar orbiter. The only problem with this scenario was that the UR-500 wasn't powerful enough to lift Korolev's LOK, so it needed an upper stage. Naturally, both Korolev and Chelomei wanted to use their own designs for this, but neither would be up to the task of lifting the full L1 design, so it was decided to simply abolish the living module from the L1. Now all that was left of the L1 was a Soyuz stripped of all creature comforts, it would have made a very uncomfortable spacecraft had anyone been asked to fly in it for a week or more. The final configuration of the lunar orbital system was called the UR-500/7K-L1. The 7K was the basic designator for all Soyuz spacecraft. The rocket would use the Blok D second stage originally designed for Korolev's N-1 so that it could be tested earlier in the program in much the same way as the Saturn IVB was test flown before being used for Apollo. This small victory for Korolev was to be his last.

On January 14th 1966 the Chief Designer, had been admitted to hospital for a routine operation to remove intestinal polyps. He was in surgery when the doctors discovered a large malignant tumor in his abdomen. His jaw had been broken in the Gulag three decades earlier and so the doctors had been forced to help him breathe through a tracheotomy. After considerable effort to remove the tumor, Korolev died due to overstraining his heart and respiratory complications. The blow to the Soviet program was incalculable. Korolev had been the driving force behind the Soviet efforts since the very beginning, and his tenacity and sense of purpose had pushed the program past innumerable obstacles over the previous three decades. OKB-1 would now fall to his right hand man, Vasiliy Mishin.

Mishin was a great supporter of Korolev's dream for the N-1. After a state funeral for Korolev, Mishin regrouped the team at OKB-1 and pushed on. The Soviet program was falling behind. The Americans had flown four successful manned Gemini flights in 1965 while all the

"Soyuz Lite"
The 7K-L1/Zond lunar orbiter

Luna 9

Soviets had to show was a string of at least ten failed planetary and lunar probes. However, less than two weeks after Korolev's death the Molniya booster finally delivered the *Luna-9* payload to a semi-soft landing on the moon. It deployed its petals and sent back the first pictures from the surface of the moon. It was another major achievement for OKB-1, and for Babakin's *Lavochkin* bureau, which had built the probe. But the Americans were close behind and placed their own lander, *Surveyor 1*, gently onto the moon only three months later. The race still seemed close, but no one in the west knew just how bogged down the Soviet program had become.

All through 1966 Mishin tried to pick up the pieces of Korolev's far-flung efforts. The design for Soyuz continued while Gemini flew another five missions with several dockings, spacewalks and even a high-flying ride with an Agena booster.

Mishin had been left with an unfinished, unmanned, Voskhod program. The next launch was three weeks after the Luna-9 landing. It was carrying two dogs again and was to conduct more long term studies on space exposure, especially in the problematic Van Allen radiation belts. It was to be the last Voskhod flight. Many more Voskhod structures would continue to be used as spy satellites under the Zenit name. In April the Molniya booster had another of its rare early successes and made *Luna-10* the first satellite to orbit the moon.

Now that Korolev was gone, Chelomei once more tried to appropriate the lunar landing program by offering a giant upgrade to his UR-500, designated the UR-700. This became another momentary distraction to the N-1/L3 being built at OKB-1 (the bureau had now been renamed *TskBEM*). The UR-700 would have offered a direct ascent to the moon, in much the same way as the many versions of the proposed *Nova* program in the United States. Its lander, the LK-700, looked almost exactly like the early designs for Apollo landers, like a CSM with legs. Meanwhile, Mishin's team had capped off 1966 with two more orbiters around the moon, another successful moon landing, *Luna-13*, and the first unmanned flight of Soyuz on November 28th. The unmanned Soyuz flight had ended with the vehicle miles off course, requiring that it be destroyed, lest it fall into foreign hands. Another flight was ordered but this one ended up with a near repeat of the terrible R-16 catastrophe of six years earlier. The launcher failed to leave the pad and then after ground crews gathered to *safe* the vehicle the new launch escape system ignited causing a fire to quickly spread to the lower stages. The fire destroyed the launch structure but it spread slowly enough that there was only one life lost.

The next unmanned Soyuz took flight only seven weeks later. Yet again the spacecraft suffered problems with navigation and guidance, which led to orientation and propellant issues. The third attempt at a Soyuz flight ended up way off course and came to an improbable end when it actually landed on an iceberg in the Aral sea, before sinking. The reason for the conversion to a submarine was a hole in the heat shield. Had a crew been aboard they would have died, because Soyuz crews would not be wearing pressure suits. This particular depressurization problem was fixed, but the bigger threat to life, the lack of redundancy, was ignored.

Despite these setbacks it was decided to push ahead and try to launch a crew. The first test version of the N-1 had finally been assembled in February of 1967 and was looming over the gigantic new lunar launch facility at Baikonur. The Soyuz, the Soviet lunar vehicle, flown twice and failed twice was now about to take its first human into space. At this time "Go-Fever" seems to have infected both the Americans and the Soviets. Just three months earlier the Americans had suffered a major setback when their own lunar vehicle had claimed the lives of three astronauts in a routine simulation. While the USA seemed to falter, the Russians ignored the warnings of their own failures and pushed on with renewed vigor. Just as the Americans had chosen to launch Apollo on a smaller booster before moving up to the Saturn V. On April 23rd 1967 the R-7 booster, equipped with an upper third stage, a complete Soyuz spacecraft and the new launch escape system, carried its first passenger into space. His name was Vladimir Komarov, one of the veterans of Voskhod 1. This first manned flight was given the official distinction of being called Soyuz 1. Incredibly, Komarov was supposed to actually dock with another Soyuz, carrying three cosmonauts, on this very first flight. However, Soyuz 2 would not leave Baikonur the following day as planned.

Almost immediately Komarov encountered problems when one of the large solar panels on his craft didn't deploy, severely cutting into his available power. This shortage of power soon put a stop to the launch of the second target craft. This was just the beginning of Komarov's problems. When it was decided to bring him home early, the de-orbit engine didn't fire. As if that wasn't bad enough, two of his orientation systems were not functioning. Komarov would have to orient the Soyuz and fire the engine manually. Once he had performed this difficult maneuver he was to be cursed with one more failure for which he had no recourse. The primary parachute system failed when the parachute got stuck in its container. The reserve chute deployed only to become entangled in the billowing primary drogue chute. Komarov had no chance and was killed when his Soyuz   Soyuz 1 launch impacted the ground. The decision to not launch Soyuz 2 actually saved the other three men's lives because the second vehicle had the same design problem with the primary parachute system. The death of Komarov not only revealed fundamental design flaws with the Soyuz but also put a final end to the race for the moon. There was simply no way to catch Apollo at this late stage. However, Mishin regrouped his team and work continued on the giant super booster and the L1 and L3 vehicles, while Chelomei continued to improve his UR-500.

In the fall of 1967 the first UR-500/7K-L1 (Proton/Soyuz) was readied for launch. On the 28th September the unmanned complex took flight but the first stage of the UR-500 failed, consigning much of the large expensive package to a toxic fireball. The L1 escape system performed its function and the lunar vehicle survived the conflagration. It had been hoped that the full L1 lunar orbital mission might still have been possible. However, von Braun was about to astound the world, because on November 9th 1967 his gigantic Saturn V booster finally took flight, carrying an unmanned Apollo capsule. Seemingly not deterred by this, a second attempt with the L1 took place on November 22nd but this time it was the second stage of the UR-500 that failed. Once again the launch escape system fired but the landing apparatus failed, revealing another flaw that might have killed a crew.

Since the Americans had not flown any astronauts for over a year there still seemed to be a glimmer of hope that the UR-500/7K-L1 might yet take a Russian around the moon before Apollo. That prospect was given a significant boost when on March 2nd 1968 Zond 4 became the first successful launch of an unmanned 7K-L1 mission. The vehicle flew around the moon on a free-return trajectory but once again its orientation system failed causing it to re-enter the Earth's atmosphere at an alarming pace and in the wrong place. It was noted that a crew could have survived, but Zond 4, an unmanned 7K-L1 Soyuz, was destroyed by the ground controllers. Six weeks later a further attempt failed during launch.

UR-500(Proton)/7K-L1 launch

In July another catastrophe happened at Baikonur when the upper stage of the UR-500 simply exploded during launch preparation. Once again the Soviets were lucky to only lose one person. In September, the fifth Soyuz L1, designated Zond 5 successfully flew around the moon. While all of these efforts with Chelomei's UR-500 and the L1 unmanned lunar Soyuz were happening, work continued apace to get the manned Soyuz back into space. On October 26th 1968 the first manned Soyuz in 19 months carried Georgi Beregovoi into low earth orbit on a mission to dock with another unmanned Soyuz. Over the preceding twelve months several important unmanned docking and rendezvous Soyuz flights had been conducted, with varying degrees of success. Now the Soviets were under the gun to keep up with the Americans, who had finally regrouped themselves after the Apollo 1 setback and flown the first manned Apollo just two weeks earlier.

Rumors were flying on both sides of the race that the other team was going to fly a moon mission before the end of the year. In fact the successive Zond (L1) flights combined with the clearly visible N-1 super booster sitting on the pad at Baikonur were not the only signals emanating from the Russian side. The Soviet leadership had been urging a manned L1 mission for late 1968 and so the intelligence reports that led to the historic "all-up" flight of Apollo 8 were indeed accurate, a manned L1 was a reality in desire if not entirely in the realms of possibility. Undoubtedly the surge of activity and flights out of Baikonur certainly must have spurred the American team to make their historic and risky decision. The very next Apollo would not only be the first manned flight of the Saturn V but it would go to the moon and do even more than what had been planned for L1, although the L1-Zonds had not been intended to enter lunar orbit, Apollo 8 certainly would.

Beregovoi's Soyuz flight was relatively problem-free and the Soviet team were reinvigorated by his successful and safe return. Another extravaganza was then planned to try and take some of the fire out of Apollo 8's parade. On January 14th and 15th two manned Soyuz vehicles were launched and hooked up in low earth orbit. It was the first time that two manned vehicles had ever docked in space, preceding the flight of Apollo 9 by two months. Two of the crew of Soyuz 5 actually space-walked over to Soyuz 4 and after a brief celebratory toast, closed the hatch and returned home in a different spacecraft to the one they had flown into space. It was a minor victory, lost in the cloud of media fuss over Apollo. Meanwhile, N-1/L3 and UR-500/L1, the two gigantic lunar programs, managed to maintain their cloak of mystery.

Despite what appeared to be the final arrival of Soyuz as a valid spacecraft, it must have seemed apparent to just about everyone that there would no longer be any way to derail the Apollo express. Soyuz 4 and 5 would be the final Soyuz flights to be launched before the race

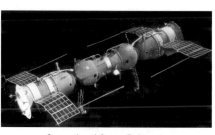

Soyuz 4 and Soyuz 5 diagram

to the moon was over, but the Soviets would not go down without a fight. Just days before the Soyuz 4/5 flight, two Venera probes had been successfully dispatched to Venus, both would enter the Venusian atmosphere in May. Another L1/Zond failed to get off the pad two days after the return of the Soyuz crews, then in February 1969 a more ambitious mission was attempted. This was to be the first of a revised version of Korolev's L2 mission, to carry a rover to the surface of the moon.

The Lunokhod moon rover

View of the Lunokhod landing stage

The *Lunokhod* was one of the great success stories of the Soviet space program. The original idea had come from Korolev and had been modified and adapted by a science bureau in Leningrad. It had originally been envisioned as a complimentary part of the manned lunar program. Korolev transferred the program to the Lavochkin bureau in spring of 1965 where another of the chief designers, Georgi Babakin completely redesigned it to now fit on Chelomei's UR-500/Proton booster, since the UR-500 was now the principal launch vehicle for the smaller lunar missions (i.e. the L1 and L2 scenarios). The rover weighed in at 756 kg, and was mounted on a lander stage of around 5000 kg. It was a large machine, over 2m wide and almost 1.5 high. It ran on eight wheels and was equipped with four television cameras. It had a top speed of nearly 200m per hour. The rovers were originally to be flown in pairs and would carry locator beacons to be used to guide a manned lander to their location. The Lunokhod (which at the time was known as the *E-8* probe) was loaded onto the UR-500/Proton rocket and on the 19th February left the pad. The booster failed at 40 seconds into the flight and another chance at a Russian first was extinguished. Then with the debris of Chelomei's Proton still scattered across the ground at the western launch facility, the time had come for Korolev's giant N-1 to finally take flight.

On February 21st 1969, fifteen months behind the Saturn V, the giant N-1 rocket, the most powerful ever built, roared upwards into the night sky above the northern launch complex at Baikonur. All of the Soviet team's dreams flew with her for all of 68.7 seconds before a giant fire erupted at the base of the first stage. Never having been given the funds to build a test stand for the gigantic Blok-A it was the first time it had been fired, and it was in full-up mode, carrying the rest of the massive rocket along for the ride. A series of unexpected design problems in the base of the first stage caused the failure and it fell back to earth to be utterly destroyed. Aboard the rocket had been a test version of the L3 lunar landing complex. Had the earlier Proton/L2 rover been successful, then the L3 would have photographed and communicated with the rover from lunar orbit. However, the double failure of the Proton/L2 and the N-1/L3 within two days of each other was a harsh blow to Mishin and all of the Russians that had worked so hard. Even then they would not give up. On June 14th, *after* the return of Apollo 10, another Proton/UR-500 carrying a lunar sample return mission blew up on launch. Nineteen days later another N-1, this time carrying an L1 orbiter designed to photograph lunar landing sites, failed only 23 seconds into the mission. The huge rocket managed to crawl about 200 m into the sky before falling back onto the launch pad and destroying it in one titanic explosion.

There was now only one minor chance left to upstage the Americans and it came on July 13th, just three days before Apollo 11's departure. On that day another of Chelomei's Proton/UR-500 boosters took flight and this time it was carrying the *Luna 15* sample return vehicle. This machine had been hastily constructed using part of the technology developed for the E-8 rover system. It represented the last gasp in the Soviet Union's attempt to wrestle some sort of victory from the continuous defeats of the last two or three years. But it was not to be. Luna 15 crashed on the moon just after Apollo 11 astronauts Armstrong and Aldrin completed their historic moon walk. This final ignominious end to the space race was kept a secret and the world was told that Luna 15 had fulfilled its planned mission. Once again the Soviets stated they had never been in the race. Their attention had been on long term space flight and their upcoming space station program would clearly prove this.

What is somewhat ironic about the space race is that the Soviet leaders were invested in embarrassing the Americans as often as they could, but the men who built the hardware were truly more interested in the long-term exploration of space and had their sights set on Mars long before they had ever been caught up in Kennedy's vortex. Although the political leaders certainly understood the significance of the Apollo achievement, they were able to simply squash it in the Soviet press and pretend it didn't happen. No one in power at that time was ever truly embarrassed by their failure because they were all long gone before the Russian people ever learned the truth about what really happened.

Back at the Soviet space design bureaus the truth hung over them like a shroud. They knew that the Americans had won the race but they had no choice but to continue on and try to regain their former glory as the world's foremost space faring nation. Plans for the N-1 continued and the Soyuz was flown again, this time in a three-ring circus. Starting on October 11th 1969 Soyuz 6 was launched followed on October 12th 1969 by Soyuz 7 and finally on October 13th 1969 by Soyuz 8. While preparations were underway for Apollo 12 the Russians flew seven people in space simultaneously. A docking was supposed to have occurred between Soyuz 7 and 8 while Soyuz 6 took the photos, but various obstacles prevented this from happening. Another L1/Zond had successfully circumnavigated the moon the previous August and several attempts continued to get sample return missions to and from the moon. At this point the Soviets began to use the (now familiar) rhetoric about robotics being cheaper and safer, but it was a thinly veiled excuse for having failed to beat Apollo. This political excuse was soon picked up and used as a mantra around the world to encourage funding robots over manned missions—a chorus that continues to this day.

The following year as America's Apollo program scrambled to recover from the near disaster of Apollo 13 the Soviets finally managed to pull off two minor victories. They landed on Venus and collected 23 minutes of data and they finally acquired their own moon rocks when Luna 16 successfully landed and retrieved 101 grams of lunar dirt and then brought it safely back to Kazakhstan. The date was September 24th 1970, just over a year too late for a political coup.

Two months later they managed to finally get the E-8/Lunokhod rover down onto the surface of the moon where it operated for an incredible eleven months, travelling over 10 km and returning more than 22,000 pictures. In many respects it was a major accomplishment because the distance traveled would be comparable to that of the American manned rovers of the following year, but, in the short term at least, for a fraction of the cost. It was an impressive display for the robot advocates who were now gaining support on both sides of the world.

During the course of the whole lunar race, Vladimir Chelomei had been methodically continuing his work on a military space station design. Chelomei, apparently more in tune to the needs of the military than Korolev, had started work on it in 1964. He called it *Almaz* and it was to have been launched by his UR-500 booster and later manned by a crew of three. Korolev had sensed another threat to his long term plans and had tried to foil Chelomei by proposing a military Soyuz derivative, but this time Chelomei would win the argument. The best part of Korolev's plans were integrated with Chelomei's. The Almaz station itself was to serve as a spying platform with a giant two meter camera aboard and so, along the lines of the American Corona system, it would have an ejectable re-entry capsule to return pictures to the ground.

The station was basically two cylindrical sections connected together and was powered by solar panels spread like wings, similar to those on Soyuz. Political expediency brought the space station program back to the forefront after the race to the moon was over. More intrigue occurred between Chelomei and the old Korolev crew, now led by Mishin. Chelomei moved from one crew vehicle design to another in his pursuit of a means of transportation to the station. Almaz was deliberately designed as a counter move to the American military manned orbiting laboratory (MOL) program which used an upgraded Titan III and a Gemini to transport the crew. Chelomei's solution involved a re-entry vehicle known as the *VA* (*Return Craft* or sometimes known mistakenly as *Merkur*) which could be launched along with a laboratory module known as *FGB* (Functional Cargo Block), the two units combined were known as *TKS* (Transport Supply Ship). *VA* looked a lot like an Apollo command module but with an enlarged parachute compartment. Access from the FGB to the VA through a hatch in the heat shield proved problematic. The VA capsule flew unmanned at least ten times on a Proton booster between 1976 and 1979 and even docked with the Salyut 6 and 7 space stations. Later the basic FGB structure was used for the smaller space station modules.

Meanwhile, Mishin, following Korolev's lead, was not prepared to relinquish an inch of ground to Chelomei and so he began his own military space station project, called Soyuz VI, that used an upgraded Soyuz spacecraft and a module called the OB-VI. But this design was not considered large enough for the higher-ups in Moscow. In fact, Ustinov specifically wanted something more substantial to counter Apollo and so he turned to Che-

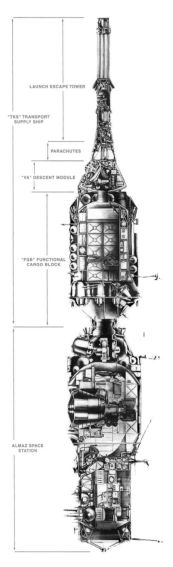

The TKS (top) and Almaz

lomei's Almaz project. Ustinov had made no bones about his dislike for Chelomei so he simply ordered the Almaz to be handed over to Mishin's bureau for completion. It was another in a long string of setbacks for Chelomei who at one point had been in control of more spacecraft construction factories than any other designer in the USSR. Now he had to sit back and watch his latest brainchild appropriated by the opposition at Mishin's TsKBEM. By early 1970 the Almaz had been partnered with another variant of Soyuz and the whole package was renamed the *Salyut*. This bizarre twist of circumstances was apparently something that neither Chelomei or Mishin wanted, but at Ustinov's behest the two design bureaus became strange bedfellows.

In the summer of 1970 the Soyuz was finally launched again, this time with two passengers. The launch took place on 1st June and the crew conducted extensive experiments during their voyage, which lasted almost 18 days, setting a new world endurance record. On their return to earth the crew seemed to be suffering from serious physical deterioration from their long stay in space, giving planners reason to pause over even longer missions.

One of the last major accomplishments of 1970 was the launch of the Soviet lunar lander atop the three stage R-7 variant now known as the Soyuz. In 1967 the USA had flown an Apollo lunar module into low earth orbit without any legs, they called it Apollo 5, now the Soviets would do the exact same thing. The LK lander with no legs was dubbed the T2K and it performed an assortment of prescribed maneuvers, exactly as planned. The date of the first two flights were November 24th 1970 and February 26th 1971.

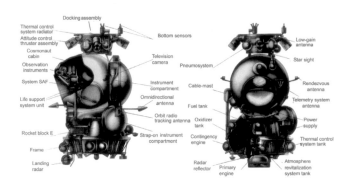

The T2K "legless" lunar lander

By April of 1971 the Salyut space station had been moved from the assembly building and was sitting atop a three stage version of Chelomei's, now relatively reliable, Proton. The station weighed almost 19,000 kg and it was 15.8 m long and as much as 4.1 m in diameter. It had four large solar arrays for pow-

er and one docking port awaiting the follow-on Soyuz crew, who were scheduled to leave four days later. On April 19th the world's first true space station
was launched into a 200 km orbit around the earth and awaited its first occupants. The crew were to fly what was to be the first of the upgraded *Soyuz-T*
model spacecraft. Unfortunately when the Soyuz 10 crew arrived it was to a
badly crippled space station that was running with only 25% of its life support
system as well as suffering from other crucial malfunctions in the scientific
equipment. Compounding the problem, the crew had some difficulty achieving
the docking and then after getting a "soft" connection it appeared they were
stuck.

Salyut 1 assembly

After involuntarily remaining docked for over five hours the problem was
solved and the crew returned home. It was an unfortunate start to an ambitious new era of Soviet space efforts but there would be more bad news
before the end of the year. On June 6th 1971 the second crew, Soyuz 11, would
leave to hook up with the Salyut 1 station. This time they were able to accomplish the rendezvous and docking and they spent 22 days aboard the hapless
station. By the time they returned to earth on June 30th, the crew of Vladislav
Volkov, Georgi Dobrovolski and Viktor Patsayev had become national celebri-
ties and had set another endurance record for the Soviets. Sadly this was to
be the darkest hour for the Soviet program. A small equalization valve in the
base of the Soyuz cabin, only about 1mm wide, had been jarred open when
two sets of explosive bolts had fired simultaneously instead of sequentially
during staging. The valve was supposed to open only on return into the atmosphere but instead it opened in space. It was too small to find and by the time

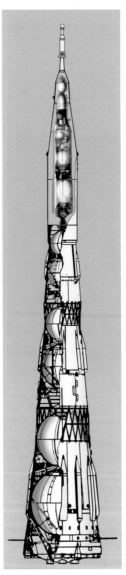

the cosmonauts had realized their peril it was too late. They died of asphyxiation before they reached the safety of the atmosphere. It was only a tiny mistake but it demonstrated just how unforgiving and merciless the space environment can be. This was a terrible blow to the Russian people, but to the teams working at TsKBEM who were distraught at the loss of their friends, it was compounded further by the fact that only three days earlier, on June 27th 1971, a third N-1 launch had been attempted. This time an unstoppable roll began in the booster just 40 seconds off the ground and the rocket literally ripped itself apart 11 seconds later. The N-1/L3 was still destined to stay earthbound. It was a grim time for the Soviet space program.

Two months later on August 12th 1971, the T2K lunar lander made its last legless, unmanned flight in low earth orbit. Once again the lander went flawlessly through its paces. All it needed was a rocket big enough to get it to the moon. It was a small victory in a summer of terrible setbacks. Soyuz would not fly again for over two years.

At the end of the year, on November 27th 1971, the Mars 2 crashed into the red planet. It could be held up as only a partial success since it had intended to soft land. A week later on December 2nd 1971 Mars 3 made a better attempt but landed in the midst of the worst planet-wide dust storm ever observed on Mars. The lander is said to have transmitted for about 20 seconds, just enough time to send one photograph, before it is believed static electricity in the dust-storm caused the delicate robot to die. On February 14th 1972 another sample return mission was successfully launched to the moon under the designation Luna 20. Then in March the Molniya/R-7 sent Venera 8 on its way to a second successful landing on Venus, this time to transmit data for about fifty minutes from the hellish surface.

⟵ N-1/L3 lunar landing system

The next major Soviet launch was on November 23rd 1972 when the full N-1/L3 complex would finally get to fly. This time all the hardware was aboard including the small one-man LK lunar lander. The launch seemed to be going fine and the booster was only seven seconds away from staging when an explosion rocked the luckless first stage and destroyed the vehicle. The launch escape system saved the crew compartment and orbiter but the fragile Russian lunar lander was consumed by fire. The reason for the failure was controversial but it seems that it was due to an unscheduled shut down of one engine. This was the end of the Soviet lunar program because only a month later the Americans shut down their own Apollo program. There was no longer any political appetite on either side for such expensive sport.

Mars 3 probe
© Courtesy Mark Wade

The next thing on the American agenda was their own space station, designated Skylab. The Soviets were clearly not going to relinquish the one thing they could call their own and so the second Salyut, which was in fact an almost unadulterated military Almaz, was launched on April 4th 1973, just six weeks before Skylab. This time the station was crippled by an explosion in the upper stage of the Proton launcher, *after* it was in space.

Now the Soviets had to sit back and watch the Americans take over the space station game. Skylab crews were seen cavorting inside the cavernous laboratory on the nightly news, while Soyuz was still undergoing an extensive refit. The heavily revised Soyuz would not take its first flight until 27th September 1973 and this time the crew would be wearing space suits. It was a tough job to modify the cabin to accommodate a crew of three, with suits, so for the time being, it was decided to make Soyuz a two man vehicle. It would be seven more years before the Soviets would feel that the machine was suitable for a crew of three again. In the meantime it would fly 27 times with a two-man crew between 1973's Soyuz 12 and 1980's three man Soyuz T3 mission.

In what now seems a bit like the tortoise and the hare, the American Skylab program burst into the limelight in 1973 and totally disappeared eight months later. Meanwhile, Soviet stamina continued to work away at fixing the Soyuz while their space station sat waiting on the ground. Five months after the Skylab crew left the US station for the last time the Salyut 3 station was launched. It was June 25th 1974 and eight days later Pavel Popovich and Yuri Artyukhin arrived to establish residency. They stayed for 15 days. Then in August, Soyuz 15, carrying a two man crew, failed to dock with the Salyut and had to abandon plans for another long duration flight, returning home after only two days.

By now it had become apparent that a long proposed joint mission with the Americans might actually occur, and so frantic preparations began at TsKBEM, which had now been turned over to, of all people, Valentin Glushko. Mishin had finally been replaced after being in charge of Korolev's empire for eight years. During his tenure the Soviets had lost two crews, lost the moon race and seen their most expensive booster disappear into the history books in clouds of flame and smoke. Mishin was not, of course, solely responsible for this, but the time had come for Glushko to finally take over his old rival's empire. In a final act of supreme irony, Glushko had relented on cryogenic engines and was now an ardent supporter of the one thing that had caused the rift between him and Korolev over a decade earlier. Glushko would now get to preside over the most important Soyuz mission ever, the one which would show the world that the Soviets were still keeping pace with the Americans.

On the 2nd December 1974 Soyuz 16 took flight in a dress rehearsal for the joint Apollo-Soyuz mission, which was slated for the following summer. Three weeks later the new Salyut 4 space station was launched and was subsequently boarded and inhabited by two crews for nearly 93 days of total occupation. After Soyuz 17 made a successful 29 day mission to the Salyut, the next designated crew were unfortunate to be the first humans to have to be ripped free of the Soyuz launch vehicle when the stages failed to separate just a few minutes into the flight. They were subjected to a 20g re-entry but the launch escape system did its job and saved their lives. It was the first ever failure of a manned launch. Their backup crew, known as Soyuz 18B, stayed on Salyut 4 for 62 days setting a new record for the Soviets, but still somewhat behind the US Skylab 4 crew.

Soyuz 19 (ASTP) launch

Finally, on July 15th 1975, Soyuz 19 took off carrying veteran space walker, Alexei Leonov and Soyuz 6 veteran Valeri Kubasov. The mission was a huge success for both the United States and the USSR with both sides scoring political victories. Soyuz 19 docked with the Apollo command module using a unique tunnel that had been carried up inside the S-IVB of the Saturn launch vehicle. Leonov and Kubasov stayed in space for a total of five days and 22 hours before returning as national heroes. In another moment of irony this would be the last manned American space flight for almost six years. While the US space shuttle was only just beginning the torturous road to its first launch, the skies were open for the Soviets.

Twenty one more manned Soyuz missions would fly in the interim between
Apollo-Soyuz and the first space shuttle flight. The Soviets would launch
ever more complex space stations but
always based around Chelomei's basic
Almaz structure. In June of 1976 Salyut
5 would be launched and would be
occupied for 65 days, followed by
Salyut 6 in September 1977, occupied
for an incredible 676 days by eleven
separate crews.

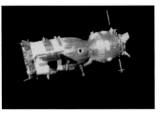

Soyuz 19 (ASTP) in orbit

While all of this was continuing a new cargo carrying version of Soyuz had
been devised at TsKBEM. It was over 7000 kg in mass and just about 8 m long,
otherwise it looked extremely similar to a standard Soyuz. It was capable of
carrying 2300 kg of cargo to the space stations in its early configurations
before being upgraded to 2500 kg. The new cargo carrier was called *Progress*
and it was first launched on January 20th 1978. This simple Soyuz derivative
would become one of the most successful and reliable spacecraft ever built.
Forty-two Progress ships flew before it was upgraded in 1989 to carry another 100kg, this later version, *Progress M*, has flown over fifty flights while a version with more fuel and less cargo, the *Progress M1*, has flown around a dozen
times, with more to come.

Flights to Salyut 6 continued until the last crew exited the station on 22nd May
1981, just a month after the first space shuttle flight. This last mission was

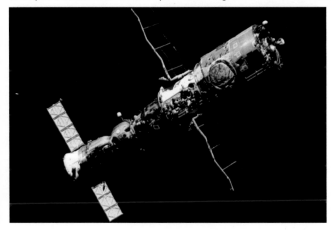

Salyut 6 with Soyuz-T docked at left

Soyuz 40 and it was also the last time that the original Soyuz configuration would fly. A temporary hiatus occurred in Soviet space flight until 19th April 1982 when the Salyut 7 station was launched. Ten crews would live aboard Salyut 7 for 812 days between May 1982 and a final excursion in March 1986. In the interim, the Soyuz-T upgrade came into its own, flying fifteen successful missions, with one more pad abort. While this long run of successful manned space flight contin-

Progress in orbit

ued, the Soviets were still up to their old tricks behind closed doors.

In 1978 Glushko had consolidated his power and was now in total control of not only TsKBEM but also his own OKB-456 engine bureau. He had also persuaded his superiors to give him Chelomei's space station and the bureau that built it. Glushko had seen the N-1/L3 super booster fail and then subsequently be cancelled in 1973, but now he would propose his own giant booster along with a moon base. This proposal was soon vetoed, being too big and expensive, but Glushko continued to push for bigger and better programs. In

1976 it had become abundantly clear to the Soviet leadership that the American shuttle was to become an instrument of the military and so it was decided, once again at the Kremlin level, that an almost identical shuttle would be built in the USSR.

Chelomei's early work on a space plane design was handed over in 1965 to yet another design bureau (the *Mikoyan* or *MiG* bureau, famous for aircraft). This research continued under the name of a program called *Spiral*. The chief manager for Spiral was Gleb Lozino-Lozinsky. The premise for Spiral was to use a hypersonic carrier aircraft as the first stage, with a total take-off mass of about 140 tons. The orbital stage was to carry one cosmonaut and was to have weighed about 10 tons. Spiral

Salyut 7 with Soyuz-T docked at bottom   was cancelled in 1969 but then briefly revived in 1972. In May of 1976 a full scale spaceplane model was built and equipped with regular jet engines. Runway and low altitude trials took place in the fall of 1977 and the prototype flew a further five times before the program was ended in 1978.

While the Spiral testing was winding down, Chelomei began work, yet again, on a 25 ton vehicle called LKS, which was to be launched on his Proton booster. Once more Chelomei's plan was to be pushed aside, this time it was a program called *Buran* which foiled his efforts. On February 12th 1976 a government decree was issued for the construction of a reusable space system. Inevitably the prime contractor would be TsKBEM, which now operated under the name *Energia*. Extensive research was done to see if there might be a better solution than that being used in the United States, but even though there seemed to be some advantage to using a lifting body wingless shape like Spiral, the more traditional winged aircraft-style shuttle seemed to be the optimum solution. The job was therefore handed over to the Mikoyan (MiG) Bureau since they had the most experience building aircraft. Mikoyan then created an entirely new sub-division, which was called *Molniya,* to handle the program. Gleb Lozino-Lozinsky was brought in from the Spiral program to oversee the project.

The Soviet shuttle would be an almost identical copy of the American design but the orbiter would not have the staggeringly powerful (and heavy) reusable main engines. Once again this was due to Glushko's lack of experience building truly powerful cryogenic rockets, especially those using LH2. Glushko also had never really had the support necessary to design and build anything like the solid rocket boosters used on the US shuttle. Therefore it was decided almost by default that the Soviet shuttle, now called Buran, would have to be launched by a new, enormously powerful, liquid booster stack. Conveniently, the infrastructure for operating such a gigantic system just happened to be lying unused at Baikonur. The massive assembly building and the equally enormous launch pads that had been built for the N-1 were lying idle and could be retooled to accommodate the Buran, exactly as pad 39 and the VAB had been in the USA.

Buran and its *Energia* launch vehicle would be the most expensive space project in Soviet history. It was originally scheduled to fly in 1979 but problems developed at every stage of the project.

The Energia/Buran
shuttle system

Unlike the American shuttle program the Buran would be the beneficiary of several unmanned orbital flights by smaller test vehicles. The unpiloted orbital rocket-plane program built upon the lessons learned from Spiral and its predecessors. Flying under the acronym BOR the vehicle looked much like an early version of the American X-38 but scaled down to about three meters in length. BOR-4 flew one suborbital flight and flew four times into orbit between June 4th 1982 and December 19th 1984. Apparently it became reliable enough that on the last two flights it was programmed to splash down into the Black Sea. These tests were used to perfect the heat shield and tiling necessary for Buran. BOR-4 was launched on one of Yangel's Kosmos C-1/3M boosters. Since BOR-4 was a different shape to the proposed Buran (it was a half scale Spiral), a further series of tests were carried out with a one-eighth scale Buran which was called BOR-5. Again BOR-5 was launched on a Kosmos from Kapustin Yar and flew suborbitally five times between July 1984 and June of 1988.

BOR-4 (top & centre)
BOR-5 (bottom)

The booster to carry the Buran was an entirely separate story. Glushko knew that he couldn't continue to rely solely on storable propellants for such an enormous vehicle and so he finally gave Kosberg the go-ahead to develop the large RD-0120 LH2 engine. After several years of struggle the design of the final booster was resolved. Since the orbiter didn't have any large engines aboard, all of the lift had to be in the main booster. It would, in a superficial way, resemble the R-7. A central core surrounded by four strap-on second stage boosters. The central core, which aerodynamically looked suspiciously like the US space shuttle's main tank, used Kosberg's 200 ton thrust RD-0120. The four strap-ons would use LOX/kerosene and would be powered by Glushko's four-chamber RD-170 engines, with one on each strap-on. The development of these strap-on boosters would ultimately lead to an entirely new family of large space launchers known, somewhat confusingly, as *Zenit* (1). This new family of boosters would be developed at Yangel's (Yuzhnoye) OKB-586, starting in the 1970's, with the first launch of the Zenit-2 in

April of 1985. In the 1990's another version, the Zenit-3SL would become the backbone of the commercial *Sea Launch* venture which provided an equatorial launch platform for satellites. Meanwhile, the mammoth, Zenit-powered, Energia stack would be capable of placing Buran into orbit or of sending 32 tons to the moon. The Energia booster stood over 58 meters tall and at its widest was 20 meters in diameter.

Delays continued to mount. The orbiter itself was too large to transport to Baikonur in one piece so it had to be dismantled and then reassembled after arrival. It would not be until 11th May 1987 that the *Energia* booster would finally fly successfully. Its payload (an anti-satellite weapon station called *Polyus*) didn't reach orbit but the huge booster was not to blame. However, although the booster was complete, the orbiter would have to wait for

Energia with the Polyus
anti satellite weapon

more than another year before it was finally readied for flight. On November 15th 1988 the fully automated Buran shuttle was placed into a 257 km orbit by the Energia booster where it remained for over two hours. It then fired its deorbit engine and re-entered successfully before cruising to a perfect landing exactly on target at the Yubileiniy air strip at Baikonur. There can be little doubt that this was an unprecedented accomplishment, made all the more poignant by the fact that it would be the one and only time the Buran would fly in space. Had the Soviet Union not collapsed, many variations of the giant Energia may have flown. The original configuration known as *Vulkan* or *Hercules* would have used eight Zenit strap-ons and an upper stage. This titanic rocket would have been capable of lifting 175 tonnes and could have lifted much of the infrastructure for a Mars or moon base in one go. A smaller version called Energia-M was built in mockup; it used two strap-ons and was designed to compete with the Proton.

While the Energia Buran program was consuming unprecedented resources another more public program had begun to take shape. After the flight of Soyuz T-14 on September of 1985 the next major launch in the Soviet manned program would be the first module for a new, much larger, space station. It was

The MIR module, principal segment of the much larger MIR space station

to be called *Mir* and the first module entered orbit on February 20th 1986, just three weeks after the catastrophic loss of the space shuttle *Challenger*. The Mir was to be the core habitation module of this new ambitious project which, when complete, was to weigh 135 tons. When it was finished it contained a pressurized volume of over 90 cubic meters. Three weeks after the Mir module arrived on station, Soyuz T-15 carrying Leonid Kizim and Vladimir Solovyev arrived to set up shop. Their mission would prove to be one of the most spectacular of any space mission ever conducted.

After spending several weeks aboard the brand new Mir module, the station and its docked Soyuz and crew were given instructions to change orbit so that they could catch up with the now dormant Salyut 7 station. This was possible because Mir's initial orbit was elliptical (172 x 301) while Salyut 7 was stabilized at 281 km in almost exactly the same plane (52 degrees). It was mostly a case of catching up, which was achieved by early May of 1986. On May 5th the two cosmonauts undocked Soyuz T-15 and piloted it across the void to a successful docking with Salyut 7. This incredible accomplishment still stands as the only time that a single spacecraft had moved from one space station to another. The crew successfully reactivated the Salyut and were able to recover experiments left behind by the last crew.

Unbelievably, Solovyev and Kizim were able to live aboard Salyut 7 for 52 days before undocking once more and returning to Mir with no less than 400 kg of equipment. They then remained onboard Mir for a further 20 days before returning to Earth. The odyssey of Soyuz T15 was without doubt one of the great voyages of the space age, demonstrating an almost brazen nonchalance as the crew skipped from one space station to another, changing orbits, ferrying equipment, performing multiple dockings and generally behaving like Buck Rogers.

Two months after their triumphant return a new revised version of Soyuz was launched and docked with the Mir module. This unmanned mission was called

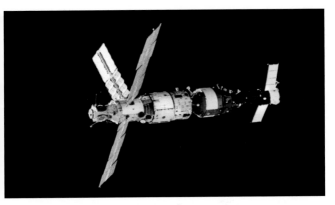

The MIR module docked with KVANT 1 module and a Soyuz at right

Soyuz TM1 and demonstrated the new vehicle's improved docking and rendezvous system. This upgraded Soyuz would be the prime spacecraft for all flights to Mir until April of 2002, with subsequent flights to the International Space Station. It would make its first manned flight on February 5th 1987 when it would deliver Yuri Romanenko and Alexander Laveikin to Mir for an unprecedented 326 and 174 days respectively. During their incredibly long tenure the two men would oversee the arrival of Mir's second module, the *Kvant 1*, which was launched on March 31st 1987 and successfully docked nine days later. Kvant would be the first adaptation of Chelomei's FGB module developed for Almaz. Crews would then permanently inhabit the Mir/Kvant station for the next two and a half years before the next module arrived. *Kvant 2* was launched on November 26th 1989 and docked ten days later. Kvant 2 was known as an expansion module. Six months later the *Kristall* expansion module arrived. The Mir space station complex would now remain fundamentally unchanged for almost five years. After Romanenko and Laveikin a further nineteen manned Soyuz TM spacecraft would dock and bring crews to Mir before yet another module was added in May of 1995. This was a remote sensing module called *Spektr*.

Since the launch of the first Mir module in 1986 to the arrival of Spektr in 1995 the political landscape back on Earth had changed immeasurably. In 1989 the presiding Soviet Premier, Mikhail Gorbachev, had finally decided to try and bring an end to the cold war tensions that had driven the space race. Working in concert with American President Ronald Reagan the era of glasnost and perestroika came and went, leaving a dismantled Soviet Union in its wake. This fundamental change in the world's power structure could not possibly pass without having profound effects on the space program. The Soviet Union fragmented back into its constituent countries. Baikonur, the principal launch site now fell within the political borders of Kazakhstan (although Kapustin Yar remained in Russian territory.) Now the old *Plesetsk* R-7 missile launch facility in northern Russia would have to be upgraded and modernized to ensure a reliable space launch capability on Russian soil.

A new era of cooperation developed between east and west and this was reflected by the first flight to the five-module Mir station by an American space shuttle on June 27th 1995. One more Soyuz crew arrived, followed by another Shuttle crew before the end of 1995. In the first three months of 1996 another shuttle and another Soyuz arrived, this time American astronaut Shannon Lucid would stay behind aboard Mir. She would stay for six months and her efforts would answer many questions about long duration space flight, some of which the Russians had still not answered to their own satisfaction. While Lucid was aboard, yet another module was dispatched from Baikonur, this one was called *Priroda* and docked on 26th April 1996. The Mir space station was an unrivalled and unprecedented success story for the Russians. Despite negative publicity from American pundits the research done aboard the giant station over its fourteen year life span has proven invaluable for future long duration space planning. The running of the station was an exercise in survivalism requiring regular docking and resupply by the Soyuz derived Progress unmanned cargo vessel. Progress frequently brought oxygen and water as well as food to the station, but the onboard systems were able to recycle thousands of litres of urine back into potable drinking water, and in as clear an illustration as you could possibly desire of the cost of putting people

Atlantis US Shuttle docked with MIR space station July 1995

into space, arriving space shuttle crews transferred bags of urine to the station to be reclaimed. The Kvant 2 module would reclaim the water by electrolysis which had the convenient by-product of oxygen, while the onboard air reclamation system was able to clear the equivalent of dozens of Progress flights worth of CO_2 scrubbers.

Despite these triumphs of endurance the station would not be without its mishaps. In 1994 an oxygen generator had caused a small fire and then in February of 1997 another much worse occurrence seriously threatened the crew's lives. The fire burned for almost a quarter of an hour and blocked the path to one of two Soyuz lifeboats, while six crew were on station. On June 25th 1997 a Progress re-supply vessel crashed into the station during what should have been a routine docking procedure. Inadequate funding and the loss of reliable sources for guidance equipment were pinpointed as the culprits. The collision with Progress really signaled the beginning of the end for the remarkable Russian outpost. Right up until the early months of 2000 efforts were made to save the Mir. Even a group of private investors tried to lease time from the Russians to use it for tourist flights, but due to lack of funding and increasing pressure from the United States government the Mir was finally de-orbited into the Pacific Ocean on March 23rd 2001. Speculation has been rampant that the decision was made in order to force the Russians to concentrate on their obligations to the International Space Station pro-

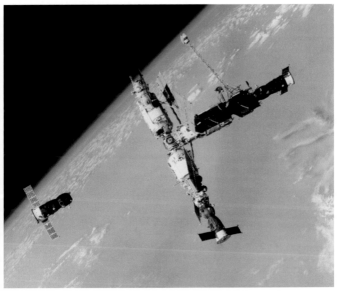

Orbital hub MIR with two Progress and one Soyuz

gram. Regardless of these stories of failure, the Mir was an unmatched success story. The International Space Station has only been in space for half the time of Mir and was under siege both politically and technically almost as soon as it was launched. The Columbia accident in which another seven astronauts lost their lives would only compound the situation, leaving ISS to be serviced almost exclusively by yet another Soyuz upgrade, the *Soyuz TMA*.

The centrepiece of the International Space Station - the Zarya module

The first ISS module was financed in the United States and built at Chelomei's design bureau, it was called *Zarya* (meaning Dawn) and weighed about 20,000 kg. Once again the Zarya was a derivative of Chelomei's FGB module developed for Almaz. It was launched on November 20th 1998 and was carried to orbit by the now steadfast and reliable Proton/UR-500. Two weeks later the space shuttle Endeavour delivered the American module known as *Unity*. The two old enemies were now firmly locked together, but the real legacy of space station hardware was still entrenched at the old Chelomei and Korolev bureaus. A few months after Mir made its spectacular descent into the Pacific Ocean the Russians launched the *Zvezda* module, a remnant of their earlier Mir program. The ISS became a three-module station on July 26th 2000 after Zvezda docked with the Zarya/Unity complex. Later that year the space shuttle delivered a docking unit and the main *P6 Truss*, which is the very large cross structure that houses many of the stations solar panels; and then in February 2001 the *Destiny* module was successfully attached. The following month the first European module took flight aboard a space shuttle. This module, named *Leonardo* was designed as a sort of moving-van. It was the first of three such planned modules, used to deliver large amounts of supplies, hardware and experiments to the station. It returned to Earth after clearing the garbage from the station. On 12th July 2001 the *Quest* airlock module was delivered to ISS by STS-104, this addition gave ISS effectively almost the same habitable volume as Mir. The following September a robotic Progress module arrived at ISS. This was another modification of the basic Soyuz structure, which by this time had proven itself as the most versatile design ever created for a spacecraft. The *Progress-M-SO* had been modified so that the usual cargo and fuel modules were replaced by a sophisticated airlock. This new addition was built at Korolev's Energia facility and once attached to the ISS it became known as *Pirs*. This small module provided extra clearance for ships docking with ISS and also doubled as a handy airlock for anyone egressing the station in Russian spacesuits.

2002 was a busy year for shuttle flights to ISS with three different missions delivering large truss structures, the S0, P1 and S1. Five more truss delivery and assembly missions were scheduled but were delayed since the only way they could be delivered to the station was by the grounded space shuttles.

Proton with ISS Krystall
module is prepared
for launch

There is a bitter irony in the long story of the Soviet/Russian space program. There can be little doubt that the colorful characters that conceived and built the world's first space program did their job despite tremendous obstacles. Political commentators in the west still find it all too easy to dismiss the Russians, accusing them of creating inferior hardware or, even worse, not caring about their human cargo as much as their western counterparts. Since the 1990's the real truth of the amazing adventure happening behind the iron curtain has finally been exposed. The Russians have had more than their fair share of hardships and grief in their long struggle to explore space. The principal players lived lives beyond the imaginations of a Shakespearean tragedy; some were tortured, some died, some fought epic battles of egotism and some were simply discarded. Clashes of personalities and political machinations, along with marginal funding ultimately demanded their price. Despite all of this there can be no denying the fact that, more often than they lost a race, they won.

When the Soviets failed, they failed spectacularly. The N-1 lunar booster was a rival to the Saturn V and it might well have succeeded had it not been cancelled at the last possible moment. Likewise, the Buran shuttle proved it could do something the American shuttle could not, it flew on its own, without pilots. It even had jet engines attached to it and was able to take off from a runway. By all accounts its structure was more durable than the American shuttle (having derived many benefits from the thousands of man-hours of research done by NASA). The Spiral program could have easily rivaled Dyna-Soar, and did in fact leave a legacy that is still being discussed today in the form of Energia's latest shuttle offering, the *Kliper*. While Vostok and Mercury stood side-by-side, it can hardly be said that Voskhod was any kind of rival to Gemini. In fact it was the Gemini program which clearly changed the status of the space race, more than any other factor, the decisions which led to Gemini and Voskhod won the race

Launch of Proton
with Zvezda

for the Americans. Comparing Skylab to Salyut and Mir is almost laughable. Skylab was an enormous and very capable facility but it was barely used before it ended up in the ocean. Salyut, Mir and now the ISS leave the Russians with a 20 year legacy of flying manned space stations. There simply isn't anything on the American side that even comes close. While Buran only flew once into space and then fell victim to the collapse of the Soviet Union, it represents one of the great unanswered questions, now that the American space shuttle has failed twice; would Buran have been an able competitor or would it have fallen victim to its sister ships' problematic design? Finally, there is the great Cinderella story of the first fifty years of space exploration, Soyuz.

Korolev's "answer-to-Apollo" has become the most versatile, durable and tenacious space faring vehicle ever built. While Apollo dashed away to the moon, returning with the garlands and laurels, Soyuz persisted. Just as Apollo lost its first crew, so did Soyuz, but out of that grim moment in 1967 came a spacecraft which continues to work like a faithful plow-horse. While Apollo lays enshrined behind glass in museums around the world, it is Soyuz that has kept the International Space Station flying and habitable. As of spring 2006, Soyuz has flown an incredible 96 manned missions and countless unmanned flights of the Progress, Zond and other variations since its birth nearly forty years ago. Two missions proved fatal and two almost so, but literally hundreds of people have seen their home planet from the tiny windows of Korolev's greatest legacy and returned home to spread the word, including the world's first space tourist, Dennis Tito, who ultimately was able to parlay his ticket to Mir into a ticket to the ISS—flying on a Soyuz.

Today the Russian space program is as Byzantine and convoluted as it was at its inception. Literally dozens of Soviet agencies and bureaus have now been reborn as competitive corporations, vying for international dollars in a tough market. Almost all of the principal players have long since left this Earth. Glushko, Korolev, Chelomei and most of their contemporaries may be gone but their vision remains. American satellites are launched on advanced Atlas missiles powered by derivatives of Glushko's engines and American astronauts fly to and from the ISS, the direct descendant of Almaz, aboard Korolev's Soyuz. Meanwhile Chelomei's Proton booster has been upgraded and since 1969 has replaced the Molniya as the primary launcher for planetary probes. The final history of humanity's space faring will likely never be written, but there can be little doubt that all future efforts will always recognize the immense contribution of Russian spacecraft.

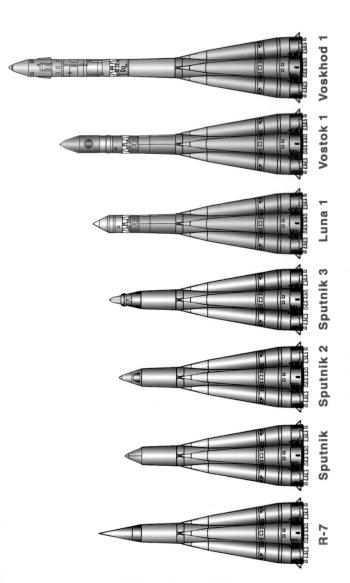

Voskhod 1

Vostok 1

Luna 1

Sputnik 3

Sputnik 2

Sputnik

R-7

R-7 derived launch vehicles

© Robert Godwin

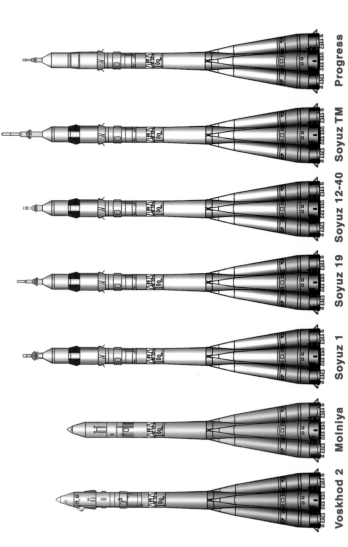

Progress

Soyuz TM

Soyuz 12-40

Soyuz 19

Soyuz 1

Molniya

Voskhod 2

R-7 derived launch vehicles

© Robert Godwin

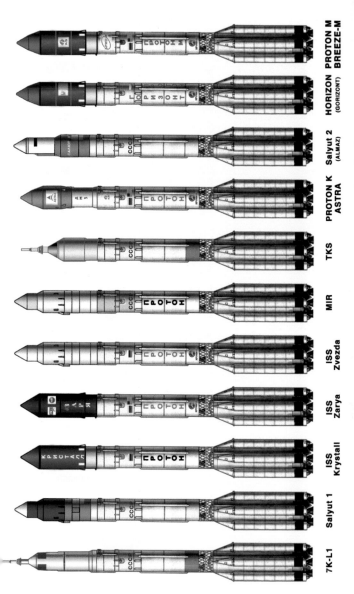

A selection of UR500/Proton configurations

© Robert Godwin

The basic R-7 ICBM
launch vehicle

Mock-up of Sputnik 1 in museum (above)

Cutaway of Sputnik 3 (Object-D)
in museum (below)

Vostok 1 carrying Yuri Gagarin
roars from the pad
April 12th 1961

Gagarin approaches the R-7/Vostok
launchpad (inset)

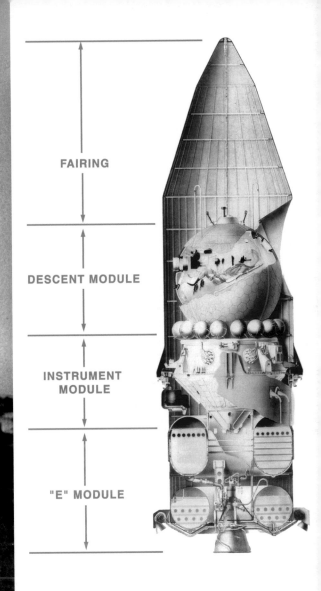

FAIRING

DESCENT MODULE

INSTRUMENT MODULE

"E" MODULE

Vostok, the first manned spacecraft flew
six times with human occupants

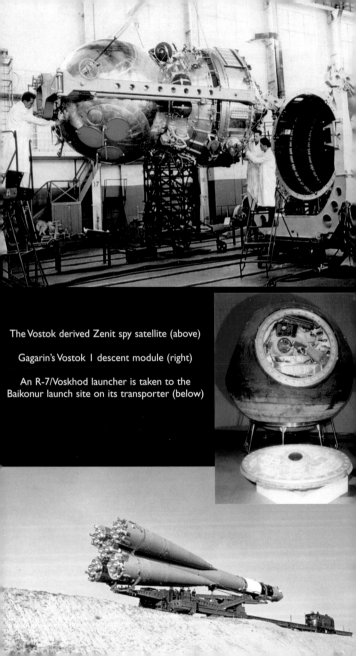

The Vostok derived Zenit spy satellite (above)

Gagarin's Vostok 1 descent module (right)

An R-7/Voskhod launcher is taken to the Baikonur launch site on its transporter (below)

Voskhod is prepared for flight (top)

The Voskhod 2 airlock (above)

The unique fairing for Voskhod 2 (left) showing the round airlock cover (left)

Voskhod 2/R-7 in the shop (left) and
on the launch pad (above)

Alexei Leonov becomes the first
spacewalker (below)

(Left) Chelomei's answer to Apollo, the "VA" descent module for TKS/Almaz, with escape tower and parachute compartment

A Molniya launcher. The four stage R-7 upgrade was first used for inter-planetary probes like Venera (right)

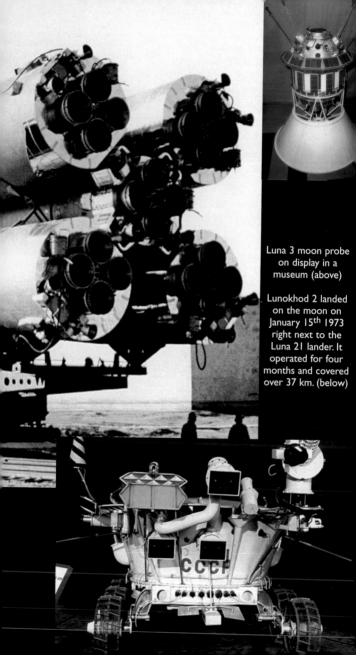

Luna 3 moon probe on display in a museum (above)

Lunokhod 2 landed on the moon on January 15th 1973 right next to the Luna 21 lander. It operated for four months and covered over 37 km. (below)

The L1 lunar complex is prepared for launch atop a Proton launch vehicle

Opposite are two pictures of the 7K-L1S an unmanned Soyuz variant which was launched atop the giant N1 rocket instead of the LOK/LK orbiter/lander combination

The giant N1 rocket (opposite) is prepared for launch. The three stages are pictured here. 1st stage with 30 engines (above) 2nd stage (at left) with eight engines and 3rd stage (below)

The full L3 lunar landing complex (opposite left) comprising the LOK orbiter and the LK lander (concealed inside)

The N1 on its erector (opposite left) The N1 first stage (left)

The first three N1 stages in the assembly building (below opposite)

N1 takes flight (below)

Salyut 4 is prepared in the assembly building (above)

The first Salyut atop a Proton rocket
is prepared for launch (top right)

Proton and Salyut are awaiting connection
in the assembly building (right)

The world's most successful spacecraft,
Soyuz in flight (right)

R-7/Soyuz launch (below)

The ASTP-Soyuz 19 descent module on
display in a museum (right)

Soyuz in the shop (left)
R-7/Progress launch (above)
R-7/Soyuz at Baikonur (below)

Spiral shuttle model (top)
Energia derived *Zenit* booster (centre)
Energia M mockup on pad (bottom)

(top to bottom)

Buran aboard Antonov transporter

Buran-Energia on erector

Buran flying with jet engines

Buran-Energia in assembly building

ia-Buran on the pad at Baikonur

Proton-Astra satellite launch (above left)
Kosmos medium lift satellite launcher (above right)

Kvant 2 module is prepared to be dispatched to
the MIR space station (below)

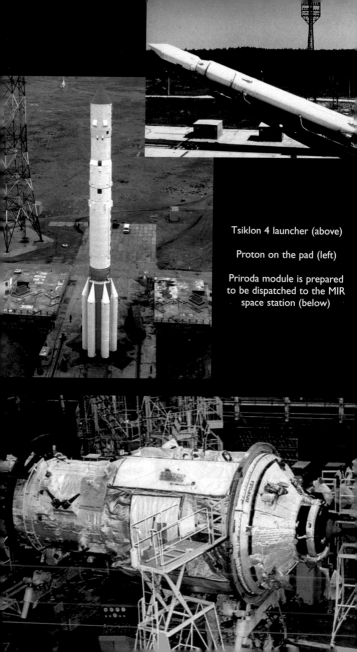

Tsiklon 4 launcher (above)

Proton on the pad (left)

Priroda module is prepared to be dispatched to the MIR space station (below)

The next generation of Russian booster is known as the *Angara*. This picture (left) shows it with the proposed Kliper mini space shuttle
© Courtesy Mark Wade.

Vladimir Chelomei's legacy lives on today in the form of the Krunichev Space Center. In cooperation with corporations like International Launch Services Krunichev sells the Proton for commercial launches. Below left is a Proton-M with the latest fourth stage known as the Breeze-M, while at right is a Proton-K carrying a Sirius digital radio satellite.